DATE DUE			
MAR – 6 1996			
MAY 13 2004			

Quelling the People

QUELLING THE PEOPLE

The Military Suppression of the Beijing Democracy Movement

TIMOTHY BROOK

New York   Oxford
OXFORD UNIVERSITY PRESS
1992

Oxford University Press

Oxford   New York   Toronto
Delhi   Bombay   Calcutta   Madras   Harachi
Kuala Lumpur   Singapore   Hong Kong   Tokyo
Nairobi   Dar es Salaam   Cape Town
Melbourne   Auckland

and associated companies in
Berlin   Ibadan

Published by Oxford University Press, Inc.,
200 Madison Avenue, New York, New York 10016

Oxford is a registered trademark of Oxford University Press

Library of Congress Cataloging-in-Publication Data
Brook, Timothy, 1951–
Quelling the people : the military suppression
of the Beijing democracy movement /
Timothy Brook.
p.   cm.   Includes bibliographical references (p.   ) and index.
ISBN 0–19–507457–2
1. China—History—Tiananmen Square Incident, 1989.
I. Title. DS779.32.B76   1992
951.05′8—dc20   92–16396

Excerpt from "Tiananmen Square" by John Simpson
is reprinted by permission of
© Granta Publications Ltd., 1989.

2 4 6 8 9 7 5 3 1

Printed in the United States of America
on acid-free paper

To Fay and her world;
for Jonah and his.

He who exults in killing people
    cannot have his way with the empire.

When many are killed,
    one should weep over them in sorrow.

LAO ZI
Dao de jing, ch. 31

Preface

The incident is now known to all. On the night of June 3, 1989, tens of thousands of soldiers armed with assault rifles forced their way into the city of Beijing and drove unarmed student protesters from the central square at Tiananmen. When hundreds of thousands of citizens and students blocked their paths, the soldiers opened fire. On the morning of June 4, thousands lay dead and dying in the streets, the hospitals, and the homes of Beijing.

I was not in China when the guns went off. I had left Nanjing, where I was researching a historical topic, a month earlier, just as the Democracy Movement was finding its feet. Like everyone else half a world away on June 4, I watched the Army's occupation of Beijing on television. I saw armored personnel carriers drive at high speed through large crowds when there was no armor to impede them. I saw soldiers equipped for battle use military tactics designed to fight an opposing army when there was no army. I saw young men in bloodied shirts carry away those who had collapsed from bullet wounds. And for days after the crowds were off the streets, I saw soldiers indulge in senseless displays of violence. The incident as it unfolded struck me then as historically and morally unambiguous.

With hindsight, it still strikes me that way. But the path between first impression and final judgment has not been a straight line. My impression that using combat weapons against unarmed civilians was a moral failure has remained unshaken for all but about five minutes of the past several years since June 1989. Historical fact has been harder to attest. As I probed the incident, precise details proved elusive. The eyewitnesses I interviewed gave conflicting accounts, frequently confusing what they saw with what they believed. Occasionally extravagant assertions by Democracy activists on one side and government propagandists on the other only complicated the task of figuring out what happened. Activists made unreasonable assertions about what the Army was trying to do that night; propagandists sought to whitewash the whole affair by simplifying the nature of the conflict and the means used to resolve it. Fact was not simply there for the taking. It had to be pursued.

So I went to Beijing. The city had been my home fifteen years earlier. During those last years of Chairman Mao's life, the twilight of the Cultural Revolution, Beijing lay in the backwater of an imperial age. Most of the city walls had come down by 1974, but a few of the old gates still stood, crumbling under the wash of sand that blew in every spring from the Gobi Desert. Inflated Stalinist architecture had ballooned around the central square at Tiananmen, but the faded vermilion walls and broken golden tiles of the Forbidden City had managed to resist both eradication and neglect. It was a city straddling a known past and an uncertain future. And, as was true of all of China in 1974, it was a city of the dignified poor. The adorned was shunned. Simple blue jackets clothed government functionaries, padded rags covered the bodies of old men consigned to a lifetime of street labor for barely remembered political crimes. Few swindlers; no beggars.

When I arrived in Beijing on the last day of September 1989, I found a city much changed. Two ring roads, plus another in the planning stage, belted the city. The sites of the old city gates were now overpasses on the 2nd Ring Road. Some of the old neighborhoods of walled compounds threaded by narrow alleyways had disappeared. Familiar shops were gone, or had been resurrected behind shiny facades and different names. Concrete towers had shot up in the suburbs. A certain kind of prosperity was cutting the city off from its past; the future looked like any Third World capital.

The people of Beijing were another story. They were much the way I remembered them. Better dressed, perhaps, and slightly less inclined to spit on the streets. But still the same rough edge, the same enthusiasm for street arguments, the same taciturn demeanor toward outsiders, the same stolid pride. People you might not warm to but could trust. Yet there was one difference, a day-and-night difference that made Beijing people in the autumn of 1989 as I had never seen them before. They wanted to talk. They wanted to tell me how they felt about everything that had happened to them and their city since the students had taken over Tiananmen Square the previous spring. Just under the surface smoldered contempt, cynicism, and anger, and it took only the mildest question for these emotions to pour out.

I got in a taxi and asked the driver to take me to an old bookstore I knew. He refused. Well, not exactly refused. He said he would take me there eventually, but first he would drive me around in circles to tell me how he felt about the Massacre and everything since. And he would not charge me for the extra time.

"We don't believe anything the government says any more," he confided. "Look at that so-called martyr Cui Guozheng, the soldier they said people murdered and strung up from the pedestrian overpass at Chongwenmen. It's all lies. He took off his uniform and went AWOL. When the Army couldn't find him, they had to hide the fact that he'd disappeared. So they got hold of a corpse from one of the hospitals and strung it up, claiming it was him. The government said hooligans did it. What hooligans? There were no hooligans out that night. Look at my car. There's not a scratch on it. They didn't attack my car, they didn't attack any of the cars in this cab company. This wasn't hooliganism. The people were battling an army."

I went to a video store to buy copies of the government's propaganda films about the Massacre. I spotted what appeared to be the titles I wanted on a shelf behind the counter and asked a store clerk to get them for me. She ignored me and walked away. Another clerk finally came over to help me. One video carried the suggestive but ambiguous title "Defenders of the Republic."

"Is that about June 4?" I asked.

"I'm not sure," was her curt reply. She was clearly not interested in selling this particular title.

"Could you put it in your VCR and let me take a look?"

As she slipped the cassette into the machine, the entire store fell silent. The first store clerk turned away.

"I don't need to see this," I overheard her say to someone else behind the counter.

The picture flashed on the screen: the earnest frown of a round-faced young soldier, red flags fluttering against a background of Tiananmen Square in sunlight, heroic orchestral score swelling. No one spoke.

"I'll take it," I said, quietly. She ejected the cassette and handed it to me. Her glance seemed to say that she at least understood why I wanted this film. It was evidence.

The videos I got that day were evidence of several different things. The footage they included was evidence of the confrontations in the streets between citizens and soldiers. But the editing was evidence of what the government wanted believed. Scenes had been selected and edited in ways designed to obscure what had happened. Visual clips were taken out of context, and their meanings changed. Real events were sliced into fictional plots. Sequences were manipulated to demonstrate that the citizens, not the soldiers, initiated the violence. The videos did not deny that civilians had been killed, but they switched the focus to the soldiers' injuries and indignities. They told of patience in the face of insult, restraint in a sea of excess. They shaped a story that justified the suppression and the means used to carry it out. The bookstores carried the same story in printed form. Arrange the facts in the right narrative and the charge of moral failure evaporates.

The day after I arrived in Beijing was October 1, National Day. The celebrations that year commemorated the fortieth anniversary of the People's Republic. Once again, real events were sliced into fictional story lines, but on a grand scale. Tiananmen Square was decked out with enormous rafts of flowers. But the Square had been under military cordon since June 4, and the people of the city were not permitted to get near the anniversary displays. Bright red banners on public buildings gloated over the success with which the Army had slammed the "turmoil" of the previous spring. There were festivities in the public parks, but they stopped at noon. People were ordered to stay home and not go out in the streets that evening. The city was closed to outsiders, and the hotels where Chinese stayed were under curfew. Riot police stood at major downtown intersections to prevent anyone from getting within two kilometers of the Square.

That evening on the Square, thousands of dancers acted out the appearance of joyful celebration for television cameras and a tiny audience of octogenarian leaders sitting above them on the old palace entrance gate of Tiananmen. The performance was followed by the most extravagant fireworks display anyone in Beijing could remember. During the celebrations, the television cameras panned to the northwest corner of the Square. Groups of students from sixteen universities, with their school banners behind them, were dancing in an awkward, impromptu style meant to suggest spontaneous joy. As long as you did not notice the absence of Beijing and Beijing Normal universities (their students had led the Movement), and if you were willing to forget that there had been killing at the northwest corner of the Square, you might have thought that the students and the government were reunited. The thought that occurred to most viewers was that these students were dancing on their comrades' graves. The word later was that they had been threatened with a month in boot camp if they refused to go, and that they were stripped and searched before being ushered into the Square. Appearance, as so often in Chinese politics, is everything.

I asked my talkative cab driver what he thought of the festivities.

"It was a way of cheating people," he declared. "Think how much it cost for the celebrations, the flowers, the fireworks. We're supposed to be on an austerity drive, and the government goes and wastes all that money. Who is it trying to fool?" He paused. "It was—how shall I describe it?— distasteful."

The same could be said of the entire propaganda campaign the Chinese government mounted inside and outside China. The idea it has projected is return to normalcy, and everything has been done to get the right images to convey this idea. Tiananmen Square was opened temporarily for foreign visitors the following day, before the potted chrysanthemums were sold off at a steep discount. As I walked across the Square, followed by the eye of an Army photographer's camera, I could see spots where the image had been touched up. The marble balustrades around the Martyrs' Monument in the center of the Square had been patched to hide the bullet damage. The burned hedges had been replanted. But the chalk marks showing the students where they were supposed to romp the night before still had not been scuffed out. And underneath the chalkings were the telltale tread marks of the armored vehicles that had swarmed into the Square on the morning of June 4.

The Chinese government's sole hope is amnesia. It particularly wants the outside world, with its coveted technology and capital, to forget the past and join in the illusion of normalcy. It asks that we succumb to its logic: That the Democracy Movement was significant only in the eyes of a tiny group of disenchanted intellectuals seduced by Western culture, egged on by an overzealous foreign press corps. That what the soldiers did, they did in self-defense against rioting mobs. That the casualties were far fewer than people imagined. That nothing really happened. That nothing has changed.

Jiang Zemin, the man who leapt to the post of Party secretary-general after the Massacre, put it in Shakespearian English for American television viewers on May 19, 1990, exactly a year after the Army first tried to enter Beijing. The international concern over the Beijing Massacre was, he said, "much ado about nothing."

I cannot agree. The Beijing Massacre was the last straw breaking the thin bridge of trust that, until June 4, still stretched between the government and its people. It has changed everything. Some months after the Massacre I predicted to a Chinese acquaintance in Canada that the future could not remain dim, and that political change would come to China in three years at most. He scoffed and said twenty. His estimate now seems the more realistic. Like most who could get abroad, he has made the bitter choice of abandoning his country. There is no longer any honor in returning home to serve China's needs; there is no hope of making a difference. In the long run, no government can withstand such alienation. But that run may be long indeed.

The purpose of this book is to establish a reasonable record of the historical incident we call the Beijing Massacre. It is also to show that the incident does matter. Relying on eyewitness testimony as well as government publications, I have tried to make sense of the ways in which the state employed violence to suppress popular support for the Democracy Movement. This is not a political analysis: no more than half a dozen members of the Communist Party Central Committee are even named in this account. Nor is it a study of the Democracy Movement. What the Movement stood for, what strategies the students took, what goals they adopted or failed to pursue—these matters are being discussed at length in activists' memoirs and scholars' studies. I leave them for others to debate. My central preoccupation here simply is to chronicle and evaluate the use of violence against civilians. It is an issue that is independent of the political claims of either side. The students' dashed hopes are troubling, but much more so for me is the means that brought them down.

If I ask that the memory of the Massacre be preserved, I do so not to banish China to the wilderness but to help clear a path out of that wilderness. However neatly the scar heals over, the wound remains. My intention is not to condemn the culture or the people of China; I have lived too long among both to do that. Rather, it is to lay before the world community the facts as I have been able to reconstruct them, and to share the conclusions to which they have led. By realizing that this ado has been about something, we may anticipate a world in which the suppression of unarmed people by military force is regarded as a solution to nothing.

Métis-sur-Mer   T.J.B
July 1991

Contents

Abbreviations

The abbreviations below are used in the text. Abbreviations that appear only in the notes are listed at the head of that section.

APC	armored personnel carrier
ARV	armored recovery vehicle
Beida	Beijing University (Beijing Daxue)
Beishida	Beijing Normal University (Beijing Shifan Daxue)
NPC	National People's Congress
PAP	People's Armed Police
PLA	People's Liberation Army
PRC	People's Republic of China
PSB	Public Security Bureau (police)
Renda	China People's University (Zhongguo Renmin Daxue)

Calendar of the Beijing Democracy Movement

Saturday	Sunday	Monday	Tuesday	Wednesday	Thursday	Friday
			April			
15	16	17	18	19	20	21
22	23	24	25	26	27	18
29	30					
			May			
		1	2	3	4	5
6	7	8	9	10	11	12
13	14	15	16	17	18	19
20	21	22	23	24	25	26
27	28	29	30	31		
			June			
					1	2
3	4	5	6	7	8	9

Chronology of the Beijing Democracy Movement

April 15: death of former Party Secretary-General Hu Yaobang

April 19: students attempt to enter Party headquarters at Zhongnanhai

April 20: students again attempt to enter, some are beaten

April 22: demonstration in Tiananmen Square during funeral for Hu Yaobang

April 25: outside troops begin to move into Beijing Military District

April 26: *People's Daily* publishes editorial condemning the students

April 27: first major street demonstration

May 4: second major street demonstration

May 13: beginning of the hunger strike in Tiananmen Square

May 15: Soviet President Mikhail Gorbachev arrives in Beijing

May 17: Zhao Ziyang's plea for toleration overruled by Deng Xiaoping; over a million march in support of the hunger strikers

May 19: first Army attempt to occupy Beijing

May 22: Army orders its soldiers to pull back to the suburbs

May 23: junior officer falls from a truck and dies during the pullout

May 26: troops begin to infiltrate to positions inside the city

May 29: Goddess of Democracy erected in the Square

June 2: final hunger strike; night troop movements discovered and stopped

June 3: second Army attempt to occupy Beijing, soldiers open fire

June 4: Army takes control of Tiananmen Square

June 5: tank fire heard at night in the suburbs

June 7: soldiers strafe the Jianguomenwai foreign diplomatic compound

June 9: Deng Xiaoping congratulates his officers on the suppression

Quelling the People

1

Introduction

As the century draws to a close, the list of human slaughter since 1900 grows prodigious. The twentieth may not have been any more vicious than any other century. It may be only that our technologies have placed in our hands weaponry more lethal than anything Julius Caesar or Genghis Khan could have dreamed of. First the Chinese "fire lance" in the tenth century, then the European arquebus in the sixteenth, then the Soviet AK-47 in the mid-twentieth, then the Type 56 assault rifle in the hands of Chinese soldiers on the morning of June 4, 1989: Prometheus is unbound. None of this evolution had to happen, but it did. It has given us a technical capacity that makes our political and moral incapacities so much more dangerous. We may be no worse than our ancestors in our flaws, but we are far beyond them in our means. Yet to excuse our atrocities by pleading our technologies betrays the century's lack of wisdom as surely as the acts of slaughter themselves.

The Beijing Massacre is one more entry in the register of our common barbarity. It was shocking to all who witnessed it at the time, but it was not inconceivable. The century has witnessed so many spectacular slaughters that the killing of a few thousand civilians in the early hours of June 4, 1989, does not fall outside what we are capable of imagining. We accept that such events happen, even as we deplore that they do. We also accept that they may make no difference to the disposition of power in the world we inhabit. It happened; it was over; and nothing in the shape of power in China seemed to change.

At least we know it happened. June 4 has not disappeared into the gray area of contested fact. Unlike other atrocities this century, the military suppression of the Democracy Movement is accepted as having occurred. Everyone agrees that soldiers slaughtered civilians. Only the number is contested, due to the Chinese government's refusal to permit an independent reckoning of casualties. But not even the government pretends that the Army did not open fire, or that civilians did not die. It denies only that the shooting deserves the status of massacre.

A massacre, to be one, needs three conditions:

- The strong must slaughter the weak.
- They must slaughter them in large numbers. (What is considered "large" depends on the context: a dozen would do on a small street, thousands might be needed on a battlefield.)
- And the strong must regard that slaughter as appropriate. Indeed, if the killing becomes a matter of public knowledge, the strong must celebrate the slaughter. They will call it a victory, or a pacification, or perhaps even the exercise of revolutionary violence. The label they put on it will override the suffering while justifying the loss. They will exult.

Only the weak will call it a massacre. The word is their weapon in an unequal contest that they, for the present, have no hope of winning. The word acknowledges the moral burden of killing that the strong will not concede. It confers dignity on an undignified experience of loss. Naming is how the weak fight back on the battlefield of language after they have lost on the battlefield of arms.

The Chinese government does not dispute the central fact of the killing. What it does dispute is the quality of the act. It does so by disputing the name. It has chosen to call the exercise of state violence on June 4 "quelling turmoil." The front end of this phrase is relatively neutral. As in English, the Chinese term for "quelling" (*pingxi*) has an archaic flavor, the sort of word that might be used to describe a military campaign in the eighteenth century. The emotive burden is on the back end. Turmoil has negative connotations in most languages. In Chinese, "turmoil" (*dongluan* before June 4, *baoluan* or "violent turmoil" thereafter) signifies a chaotic condition that works against the good order of society. Nothing good can come out of turmoil; life can proceed only when turmoil subsides. Quelling is thus an appropriate way of dealing with something so irredeemably bad. Suppression becomes positive when the object of suppression is negative. The government's logic is simple: despite the deaths, there was no massacre.

Among the people of Beijing, however, the word on everyone's lips on the morning of June 4 was not "turmoil" but "massacre" (*tusha*). No other word could do justice to the experience of the night before.

My research on the military suppression of the Democracy Movement in Beijing has led me to defend the claims of the people against the government, of the weak against the strong. On the basis of interviews with dozens of eyewitnesses, supplemented by a careful reading of student leaflets, activists' memoirs, and official publications, I have come to a conclusion that is unambiguous: there was no turmoil in Beijing until the Chinese government created it. The military violence unleashed in the streets of the capital did more to heighten turmoil than to quell it. What the Army performed was not a limited security operation appropriate to a civilian situation but a full-blown military operation designed for war. When the Army used full force to pacify the people, staggering casualties were unavoidable. This was a massacre.

The shock in Beijing reverberated around the world. It was not only the gruesome spectacle of violent death that shocked. It was also the shock of surprise. The world was surprised that China, having so recently claimed its place in the community of nations, would resort to public bloodletting to resolve a crisis. The indiscriminate shooting of civilians was not what we expected of China. Wasn't this the venerable civilization where men of literary, not military, skills traditionally ruled? Wasn't this the revolutionary society dedicated to ending oppression? Wasn't this also the newly modernizing nation committed to meeting international standards of political and economic conduct?

The Chinese people themselves thought so. A decade of economic reforms and slow political liberalization through the 1980s induced them to believe that the harsh politics of the 1950s and 1960s—the liquidation of the landlords, the Hundred Flowers campaign, the antirightist purges, the Great Leap Forward, the Cultural Revolution—were nothing but faded nightmares from the past. And anyone in China could have told you that none of these campaigns was brought to conclusion by ordering soldiers to fire in the streets. Once during the Cultural Revolution, a local commander in Guizhou province ordered his troops to fire on Red Guards, but it happened only once, without central authority, and the commander was disciplined for his action. Who in 1989 could imagine that violence would be brought to bear this time? In a subsequent interview, astrophysicist Fang Lizhi, China's leading human-rights advocate, told me he had expected that the government would use the Army to bring the Democracy Movement to heel, but even he was surprised how it was done. "I never imagined that they would open fire on a large scale. Most certainly they could have used other means. There was no need to kill people in order to restore stability. One might expect that one or two people would be killed accidentally when the Army advanced. This sort of thing is unavoidable. But to kill people in such numbers? That could have been avoided."

How did the killing come about? The Chinese government's version of the spring of 1989 provides no answers. Its mission has been to obscure the thinking that led to this outcome and construct a new logic that blames the victims. Without access to reliable information about how decisions were made at the top, it is difficult to challenge the authority of the official version. What is needed, then, is an alternative logic to explain the process that led to the killing. I believe that the only way to construct such an alternative logic is to analyze observable military actions. What the Army did reveals a great deal about what the Party wanted to do.

To reconstruct the logic of events, I started with what I consider the central fact of June 4, a fact beyond interpretation: assault troops slaughtered civilians. From that central fact, I worked my way back through a series of eight sequential steps leading up to the slaughter. Each step down my logical staircase brought a new discovery.

Making the first step back from the killing was simple. The reason soldiers killed civilians is that they were given the order to shoot. The decision to shoot was not a military one. Officers in the field requested permission to open fire, but the order originated with the central political leadership.

Why were the soldiers given the order to shoot? The second step back in the logic of events leading to the Massacre was also simple, at least on the surface. Soldiers got the order to open fire because their advance was stymied. The Army would not have achieved its objective of securing Tiananmen Square before dawn without it. Resistance led to impasse, and impasse led to violence. This neat connection does not mean, however, that violence was unavoidable. Under other circumstances, resistance would never have occurred. The Chinese government would have us believe that it had no other choice. This was not an assumption that I could share.

Accordingly, for my third step, I took a different logical stairway. The reason the soldiers got stopped in the streets was not simply because the people resisted them. They got stopped because they were part of an operation that was badly executed and poorly coordinated. Their own incompetence fed the confusion that led to violence. Invading columns had no backup plans should they fail to meet their objectives. Radios did not work. Trucks broke down. Food ran out. Some APC commanders had no maps of the city. Riot squads fired tear gas into the wind that blew back in their faces. Units came in at the wrong times and places, stumbled over each other (resulting in military casualties), and arrived at their tactical destinations at the wrong times. Their conduct aroused angry crowd responses. These responses neither they, nor their commanders, could control. All that was coordinated were the first orders given on the afternoon of June 3 to get soldiers into position, and the last predawn orders on June 4 to clear the Square. Everything in between was left to the skill and wisdom of the troops; not, in this case, a recipe for success.

My fourth step in tracing back the logic of events leading to the Massacre came with trying to understand the mess the Army made of the operation. Partly it was due to the low quality of Chinese soldiers and officers, but the principal reason for the chaos was wrong strategy. The multipronged invasion of Beijing by troops armed with lethal assault weapons was not a plan suited to gaining control of a city of eleven million hostile residents with a minimum of casualties.

Why was this kind of operation put into action? The fifth step back took me to the problem of China's military resources. The wrong strategy was adopted because the right resources were unavailable. The troops dispatched to clear the students from Tiananmen Square were not trained for the task. Most were rural boys who had never walked through a city, let alone rehearsed combat in one. Some were fresh recruits; one company was made up of teenagers inducted into the Army only forty-three days before the attack. Even experienced soldiers seemed to have little clear idea of what they were supposed to do. Add to this high levels of anxiety and fear that interfered with field conduct and with the soldiers' ability to understand and carry out orders. Worst of all, these men were improperly equipped for the job at hand. The Chinese version of the AK-47 is a fully automatic assault rifle designed to inflict severe bullet wounds on large numbers of combatants within a range of almost half a kilometer. It is not a tool for dispersing unarmed demonstrators. The strategists formulated their plan of attack based on the wrong kind of soldiers using the wrong weapons. Right there the formula for violent disaster was set.

My sixth step was to ask why large numbers of undertrained and improperly armed soldiers were sent into a situation for which they had neither the tools nor the skills. In a press interview on July 1, 1989, Premier Li Peng gave his reason: China lacked sufficient supplies of tear gas, rubber bullets, and riot control gear. No other means were available, he declared. It was merely a technical problem. His explanation takes us down another wrong logical stairway. Large numbers of soldiers were armed with AK-47s and dispatched into Beijing not *because of* a lack of appropriate tools but because of the refusal of citizens to obey the government. To government leaders, the scale of resistance seemed so vast as to require a full military response.

The next step back in my reconstruction was unambiguous. The reason citizens poured into the streets in defiance of government orders was that the government sent in the Army in the first place.

The final step was the simplest, if least easily defined. The reason the government turned to the Army was that it had lost the allegiance of the citizenry of Beijing and could think of no better means of trying to get it back. Faced with the prospect of military occupation, the citizens turned a vague support for the students into a commitment to protect them. Official spokesmen have since argued that the Chinese government had no choice but to turn to the Army to restore order. But its conundrum could be put another way: government leaders used soldiers because they could not imagine any other way of reestablishing their authority. They could understand popular opposition only as a life-or-death, you-or-me contest. The leadership faced a political problem, and chose to resolve it through military means.

The assumption underlying this eight-step reconstruction is that military action is always inappropriate for handling a political problem. If the Army's conduct in Beijing in the spring of 1989 demonstrates anything, it is that military force should not have been used. There were other ways of dealing with the problems kicked up by the Democracy Movement. Slaughter is never simply unnecessary; it is an enormous moral failure. This is the burden the present leadership must bear.

We are left at the bottom of my logical stairwell, however, with a final, perhaps unanswerable, question. Why did the Chinese government adopt a military solution to a political problem?

The reason for that choice lies partly in the contingent ways in which government leaders perceived the students' challenge. But it lies at a deeper level as well. For a historian of imperial China, the temptation to see in the events of spring 1989 parallels with the past is strong. I resist it whenever possible. China changes. So too does the world around China. Still, long-standing cultural traditions continue to color the way people think. They are part of what makes the everyday thinking of Chinese people different from the everyday thinking of people elsewhere. They are also part of what conditioned how both leaders and citizens reacted to violence.

Of course, the Chinese cultural framework cannot explain everything. Killing unarmed civilians with lethal assault weapons is hardly unique to China in this century, even in that particular season of the year. Troops in the Somali capital of Mogadishu did in July 1989 what Chinese troops were doing in Beijing in June, and with identical weapons. It may seem odd to compare China to Somalia, given the disparity between their histories and material resources, but it is not a trivial comparison. Looked at from a world context, the Beijing Massacre is a depressingly modern and universal phenomenon. Silencing civilian protest with guns is a hallmark of Third World state-building and forced economic development at the close of the twentieth century. The Chinese government is not unique in using force as its final refuge.

But it might help to begin this inquiry by thinking about some of the ways in which China *is* unique. In particular, it is worth reflecting on certain norms regarding the use of force that cannot be reduced to a universal Third World predicament. These norms may help the reader better imagine the world in which the Beijing Massacre took place.

Few cultures approve of organized violence, but almost every culture accepts its use under certain circumstances. For two and a half millennia, Chinese civilization has disparaged the ruler who resorted to military force, but it has done so not from any pacifistic impulse. Early Chinese philosophers and strategists understood that the use of force carried with it unintended, and uncontrollable, consequences. (During the tumultuous Warring States period in the two centuries leading up to China's unification in 221 B.C., an era when most of China's classical wisdom was formulated, the philosopher and the strategist were often the same person.) Force might sometimes be needed, but far better to use nonviolent, and hence more reliable, methods of controlling opposition. The risks of war are enormous because violence never traces a steady course.

Soldiering was not the honorable career in China that it was in medieval Europe. In imperial China, the ladder of success went up through the civil bureaucracy, not the military ranks. A few men over two millennia succeeded in seizing the throne by military means, but they could not hope to govern a territory so broad and a population so vast by force of arms alone. It was expertise in scholarship and administration that Chinese rulers rewarded, not martial valor. The throne still needed armies, but little status went to army officers, and none to the rank and file. A capable general excited only suspicion on the part of his emperor. Among the people too, soldiering was viewed with disdain. It offered permanent poverty and an early death. Only the poorest and most desperate drifted into the army. As the old saying went, you don't make nails out of good iron, and you don't make soldiers out of good men.

If you couldn't rule the empire from horseback, to turn an old adage around, you still had to capture it from horseback. On their ascent to power during the second quarter of the twentieth century, the leaders of the Chinese Communist Party, especially Mao Zedong, grasped this necessity. The power of the Party had to rely on military capability, for without it the Party could achieve nothing against warlords, the Nationalist armies of Chiang Kai-shek, or the military might of Japan. Only by means of an army—the Red Army, forerunner of today's People's Liberation Army (PLA)—could the revolution come to power.

The Communist leaders knew they needed an army, but they also realized that armies achieve nothing lasting without the support of the people. Their stroke of genius was to bring together the two elements of military capacity and popular support to create a peasant army that worked with and for the people. By infusing the common soldier with revolutionary spirit, his reputation, at an all-time low during the warlord era of the 1920s, was miraculously elevated. Soldiers of the Red Army were not the "bad iron" of the warlords, and they brought the Communist revolution to the peasants from whom they had come. They overwrote the old imperial images of death and destruction with a new vision of commitment to building an equitable socialist state. Through the long civil war against Chiang Kaishek and the guerrilla struggle against the Japanese occupation forces, the Red Army became a true people's army of liberation. To use a favorite phrase of Mao's, the Red Army was the fish and the people were the water. In 1949, the Army dutifully delivered China into the hands of the Party.

The People's Liberation Army entered the post-1949 era with a high reputation for selfless devotion to the nation. Soldiers were now honorable men, and the chance to serve in the PLA was every lad's dream. Victory did not lead to professionalization. Unlike other Third World armies, the PLA continued to foster integral links with society. Soldiers farmed, opened new lands for cultivation, constructed canals and railroads, operated mines and factories, undertook educational projects, wrote fiction and published newspapers, mounted plays and produced movies (at the August First Film Studios, named after Army Day, in the southwestern suburbs of Beijing). The Army did not become a military caste but remained a broad organization that contributed in highly visible ways to the social and cultural life of the People's Republic.

The PLA's high standing in the eyes of all within the new order did not prepare the citizens of Beijing for what happened when the soldiers were ordered to open fire on the night of June 3. Indeed, it led them to expect the opposite. No one was more surprised by the live fire than the people themselves. The PLA had used automatic weapons to enforce martial law in Lhasa not three months earlier, killing over 450 Tibetans. That might have warned the citizens of Beijing that the Army could not be relied on to withhold fire when soldiers were mobilized to quell internal dissent. But the episode in Tibet did nothing to change expectations of the Army among Han Chinese. (After all, the PLA had fired on Tibetans before, and most Han Chinese had remained indifferent.) The ethnic majority continued to believe in a sacred compact between itself and the PLA.

To assume honorable conduct on the part of the Army, as most of the protesters did, was to look at the wrong tradition. The suppression of the Democracy Movement would end up conforming less to principles established in China's recent military history than to traditions rooted in the vicious conflicts of the Warring States period. It was in those violent centuries before the First Emperor unified China that the Chinese conception of war took form, a conception foreign to Western notions.

In the West, war has been regarded as an occasion for the warrior to display his valor. Warfare is an exercise of honor, and the champion is he who stands brave in the face of danger and fights his opponent fairly. These notions come to us from the aristocratic traditions of medieval Europe, where contests of individual knights could decide the outcome of opposing armies. Warfare was principally a means to decide conflicts among aristocrats. The advent of impersonal arms that strike at a distance, like crossbows in the eleventh century, and guns later, stripped all real honor from the waging of war in the West. (Medieval aristocrats disdained crossbows as "un-Christian" weapons because crossbows pierced not only their armor but the illusion of their invincibility at the hands of commoners.) Yet still in Western thinking, the idea of the fair fight persists.

In China, the crossbow was already a common assault weapon by the fourth century B.C. It propelled war toward a pitch of violence no civilization had ever before experienced; it made the older tradition of individual combat obsolete in China long before Europe outgrew it. Battles in which two vast armies of equal striking power faced each other at a distance could no longer be won by individual heroics. What was needed now were ruses, as most Warring States philosophers agreed. War is a dangerous encounter and the stakes are high, they argued, so winning is more important than how you play the game. If you have to fight, it is not advisable to do so honorably, or well, or by the rules, or in a fair manner. There is no honor in fighting. To confuse honor with warfare is a peculiarity of the medieval European mind. He who insists on a fair fight is a fool because he gives himself only an even chance to win. The challenge is in outfoxing an opponent rather than beating him down: to win without ever having to join battle at all. It is not valor or honor but deception and craft that make the difference between victory and defeat.

What is respected in war is not skill in wielding a weapon. That you leave to men too stupid to win by any other means. The real skill in battle is the cunning that invisibly arranges the outcome. Even before the battle has started, the enemy's defeat is assured. Victory is not something to be wrested with a mailed fist; it should fall lightly into your hands. Warfare is a battle of wits, and the commander who substitutes brawn for craft is sure to lose. We find this ancient way of thinking in a popular Warring States treatise entitled *The Art of War*. (Less than a century later, the political theorist Han Fei remarked that every home had a copy of this book; it remains common reading today.) The author of the treatise, Sun Wu, understands that success in battle depends entirely on taking full advantage of the enemy's weaknesses while converting your own shortcomings into strengths. The keys are careful planning and deception. The book opens with memorable advice:

> Warfare is the Way of Deception.
> Hence,
>> when capable of attack, appear incapable,
>>> when using your troops, appear inactive . . .
>> seduce your enemy by conceding advantage,
>>> crush him by feigning disorder . . .
>> use his anger to irritate him,
>>> use your weak position to make him arrogant . . .
>> attack him where he is unprepared,
>>> appear where he does not expect you.

The PLA strategists of the Beijing Massacre were familiar with Master Sun's ideas. The attempts to take Tiananmen Square invoked the "Way of Deception" by employing simple ruses intended to take the students and citizens by surprise:

- Sabotage nonviolence by planting undercover agents in the crowd to channel popular discontent into a call for blood ("use his anger to irritate him").
- Offer the people combat weapons under the guise of supporting them and coming to their aid ("seduce your enemy by conceding advantage").
- Send out squads of plainclothes officers disguised as ordinary workers and street toughs to free soldiers that real workers have pinned down ("when using your troops, appear inactive").
- Make it easy for protesters to torch military vehicles ("when capable of attack, appear incapable").
- Expose some soldiers to attack and let the people take down a few easy targets ("use your weak position to make him arrogant").
- And cover it all with the confusion that is integral to a night attack ("crush him by feigning disorder").

The real surprise, of course, was not the Army's invasion but the people's resistance. The people were protected so long as they stayed within the magic circle of nonviolence, and so the Army had to use both deception and violence to goad the people into attacks against soldiers. Only by forcing the people to fight on the Army's terms—and without the Army's weapons—could the people be defeated. Once both sides turned to violent conflict, it would be a battle that the unarmed could only lose.

As much as ancient traditions can tell us about Chinese thinking regarding military actions, we cannot reduce the events of June 3–4 to devilish Oriental scheming. Curiously enough, many Chinese, and China watchers, do. They assume that a Machiavellian master plan was unfolding with cool precision at every stage of the assault. Steeped in the very martial lore I have outlined, they see sinister plots beneath the surface of every Army action. Nothing was simply what it appeared to be.

This view of June 4 as a grand orchestration credits the PLA with far more craft and capability than it deserves. Lives were lost in great numbers, and the dignity of the citizens of Beijing was powerfully affronted. Without doubt, both were direct consequences of the Army's actions, but not everything that occurred was the result of Army craft. Scheming and deception were at work, but so were mismanagement, incompetence, and unprofessional conduct. At many points, it is hard to tell the deceptive sleights from the egregious bungling. My own sense, after sifting through the evidence for two years, is that error and naked aggression had far more to do with the atrocities of June 4 than deception. Too many incidents were pure blunder, not tricks from *The Art of War*.

Error and deception, whatever their proportions in the mix, proved to be a lethal compound when combined with an apparent indifference to casualties. When soldiers blasted apart the bodies of unarmed civilians with weapons designed for military combat, they precipitated crowd reactions that produced great loss of life on both sides. At the same time, by forcing soldiers into situations that they were neither trained nor equipped to handle, by forcing them to participate in slaughter, the Army brutalized its own men. Terrified soldiers were exposed and lynched. In a televised conference with his military commanders on June 9, supreme leader Deng Xiaoping dared to assert that the soldiers who died "met death and sacrificed themselves with generosity and without fear," but it was their commanders who ordered the sacrifice.

Ancient strategists recognized that the shift to violence may produce quick success, but provides no strategy for governing in the longer term. As early as the fourth century B.C., the philosopher Xun Kuang argued this point in a debate at court against a disciple of Sun Wu. "You speak of the value of plots and advantageous circumstances," Xun Kuang observed to his debating partner, "of moving by sudden attack and stealth—but these are matters appropriate only to one of the lesser lords." Deception, he insisted, is not part of "the intentions of a true king." Master Sun's advice is fine for petty tyrants, but in the end, Xun Kuang warned, "what is really essential in military undertakings is to be good at winning the support of the people." Mao Zedong would have understood; indeed, he adapted the image of the fish and the water from Xun Kuang's metaphor of the ruler as a boat riding on the sea of the people. To rely on warfare for ultimate victory, said Xun, is "like trying to break a rock by throwing eggs at it."

In the short term, Deng Xiaoping did manage to break the rock of popular solidarity against his government. In the long term, real authority has slipped from the grasp of the Party leadership. The author of *The Art of War* understood this. For all his advice about deception, Sun Wu stressed that violence should be exercised only to preserve peace and human life. Machiavellian in his strategies, Sun was not indifferent to the fate of the people at the hands of tyrants. In the twelfth chapter of *The Art of War*, he warns that the ruler who wastes human life tempts his own demise:

> The ruler should not put his army in the field out of anger,
>   the general should not send his soldiers into battle out of pique.
> Anger can turn back into happiness
>   and vexation into joy,
>   but a country once destroyed can never be revived,
>   nor can the dead be brought back to life.
> Hence the wise ruler is cautious,
>   the good general vigilant.
> This is the Way to keep the country at peace and the army intact.

Sun Wu was not the only Warring States thinker to come to this judgment. Lao Zi, the shadowy formulator of the Daoist notions of spontaneity and inaction, accepted that the ruler must vanquish all opposition; otherwise, the ruler could not secure his rule. But Lao Zi warned against the folly of resorting to violence to attain this end. Violence is only a short-term strategy. It has no efficacy in the long run. Being an overt interven-

tion in human affairs, a rejection of consensus, warfare is "liable to rebound." Whatever a ruler achieves through military victory will be lost the
moment he has lost the hearts of his people. "Arms are implements of ill
omen," Lao Zi warns, "not the implements of the Gentleman." They
never achieve what they are called upon to achieve. When violence is
unleashed, it is not only the enemy who falls. Lao Zi phrases the power of
unanticipated consequences with characteristic felicity: "Turning back is
how the Way moves."

The men who ordered the Army into Beijing chose not to listen to Lao
Zi's advice (which appears as the epigraph to this book) to weep rather than
exult when many are killed. Deng Xiaoping was exultant. "The PLA
losses were great," he conceded in his June 9 television remarks, "but this
enabled us to win the support of the people." Not a glimmer of doubt or
remorse about the wisdom of military action. "This was a test," he insisted, "and we passed."

If the Party leadership needed a demonstration of Lao Zi's observation
that harsh actions produce contrary consequences, they had only to look
that autumn to Eastern Europe. The PLA brought the Democacy Movement to an end in China, but in Eastern Europe, where six Communist
regimes would collapse in the next six months, the Beijing Massacre produced its opposite effect. None of these regimes was able to rescue itself by
invoking "the China solution," as Eastern Europeans called the military
suppression of democracy activists. Implemented once in Beijing, the
China solution became unworkable among Eastern European leaders.
Only Nikolai Ceaucescu dared order his troops to fire on the Romanian
people that winter, and his bid to save his regime failed. (The Chinese
government reacted to Ceaucescu's execution in December by putting the
People's Armed Police on alert.) An attempt to use the army to block the
process of democratization in the Soviet Union in August 1991 met with
the same fierce public response, and the same failure.

Since June 1989, China has continued to pursue its own course. And
life goes on. Like the rest of us, the Chinese people cannot be expected to
remain in permanent shock over the Massacre. There are always other
challenges to face. China's history has been a ceaseless working-out of new
political solutions to the perennial problems of subsistence and order, and
that process has continued on past June 4, 1989. Even so, it does no good
simply to forget what happened, as several of my colleagues in the China
studies field have advised. There is no "business as usual." The Beijing
Massacre may have been only a brief moment in China's longer search for
solutions, and many pre–June 4 policies may now be back in place. But the
shooting continues to taint the Chinese government. That stain will not
fade with time. Everything for now—every policy, every platitude, every
project—is temporary, a delicate tower waiting for the wind to blow.

The suppression did not bring the history of contemporary China to an
end. Nor, for that matter, was the Democracy Movement the final word

on how China will move forward. We must be prepared to accept that political change may not come to China through sudden, radical events like the Democracy Movement. We must bear in mind too that the urge toward a more pluralistic order may not result in the political forms democracy has assumed in the West. But neither observation can take away from the remarkable heroism and disturbing violence that marked the spring of 1989. Short of a full and honest accounting of what happened during that spring, the suppression will haunt China for the rest of the century, perhaps beyond. The Chinese government must come to terms with what happened. Until it does, it cannot hope to make anything better come about. As long as it denies its past, China has no future.

2

Changing Fate
(April 15–May 19)

At 10:00 A.M. on the morning of Saturday, May 20, 1989, the city of Beijing came under martial law. The order of the State Council invoking this measure observed that Beijing was in "serious turmoil." The turmoil, it insisted, not only threatened public property and state organs but placed even the lives and property of private citizens in jeopardy. It had to stop. Martial Law Order No. 1 granted police and soldiers "the right to use every possible means" to stop it.

Never before in the forty years of Communist rule had China put its citizens and security forces in quite this situation. No major city in China had ever experienced such a sanction, with the exception of the Tibetan capital of Lhasa—but in the Autonomous Regions, different rules apply. Beijing is in the China of the Chinese; it is the nation's capital.

Martial law is an extraordinary measure. It is usually a desperate one as well. It signals that public order has come under siege; more to the point, that the authority of the government is under siege as well. This was a novel situation for China, and for its leaders.

Had martial law exploded on them unheralded, the citizens of Beijing would not have known what to expect. Martial law was what happened in other places, not at home. You read about it in newspaper reports on regimes elsewhere in the world: helicopters hovering over downtown avenues, tanks rumbling in, soldiers with rifles at the ready shooting anyone on the slightest pretext. The image of such repression was foreign. But the people did not have time to imagine martial law, to fit the extraordinary measure with their extraordinary circumstances. They found themselves in it even before it was declared. Fourteen hours before the order came down, the troops sent in to enforce martial law found blockades at every entrance leading into the city. When martial law began at 10:00 A.M. Saturday morning, nothing happened.

The government's decision to call in the military was a fateful one. It closed the first link in a chain of events that would lead fifteen days later to the Beijing Massacre. This was not a chain that had to be forged. Giving the martial law order did not necessarily mean that a massacre had to result. But armies are armies. To mobilize them is to open a chasm that might not be closed without blood. Starting in mid-April, every decision on the government's side brought the possibility of such an outcome closer. To understand the Massacre, we need to understand how Beijing came to find itself under martial law. To comprehend that, we need to go back to the beginning of the Democracy Movement and see how the government's reactions, inch by inch, brought Beijing closer to the chasm's edge.

The Democracy Movement was a precarious project right from the start. By daring to speak for freedoms not recognized in practice by their government, student activists were stepping into a black void. They did not know how far they might fall, and they took the awful chance that nothing would catch them before they hit the bottom. "The students knew it was a terrific risk," a foreign university teacher observed to me later. "But they did not want to repeat what had happened to every such movement in the past: that it would quiet down, that the government would make sure they paid a price, that the world wouldn't know. Once the Movement got going, the students could not afford to cool things down for the sake of compromise, for they already knew they would have to pay a terrible price. They wanted the world to know."

The spring of 1989 was not the first time students dared to challenge the dispensations fate handed to them. The fate of falling under Western domination in the nineteenth century, with its guns and opium, inspired Chinese students in the 1890s to believe they might turn the weapons of the West against the imperialists and put China back on its feet. But the first irreversible step came on May 4, 1919. That spring, Germany's victors met at Versailles to arrange among themselves the postwar world. China was represented, having contributed to the allied war effort by dispatching several hundred thousand laborers to work behind the lines in France, but China was not one of the arrangers. Behind the backs of the Chinese delegates, the European powers decided that Japan should be allowed to keep German colonies in China that it had seized in 1915 under the pretext of going to war against Germany.

Students from thirteen colleges in Beijing gathered at Beijing University on May 4, 1919, to discuss a course of action. What could be done about their government's unpopular decision to cave in to the Treaty of Versailles? Naively, they expected the Western democracies to respect the sovereignty of their nation rather than indulge in Great Power scheming. Determined to express their frustration at their government's capitulation, about three thousand Beijing students assembled in the large forecourt of the Forbidden City directly north of Tiananmen, the Gate of Heavenly Peace, and marched on the foreign legations nearby. When police blocked their way to the diplomatic quarter, they redirected their march toward the home of a government minister who had negotiated loans from the Japanese, and set it on fire. In clashes with police, one student died and another thirty-two were arrested. Public sympathy for the students was so overwhelming that the police were obliged to let them go. Emboldened by their newfound power, the students formed a citywide union and agitated in favor of reform. At the core of their demands was the call for democracy and science to replace the worn-out, autocratic traditions of the past. The ferment came to be known as the May Fourth Movement. Out of it too arose a vibrant group of young intellectuals determined to create a new world for China and win for themselves the role of conscience of the nation. Out of it arose as well the Chinese Communist Party.

Seventy years later, on May 4, 1989, the students of Beijing were once again at Tiananmen. This time they gathered south of the gate on the large ceremonial space that had not existed in 1919. The police lines put up to control them collapsed. No one was beaten, no one arrested, although the students in 1989 were challenging a vastly more powerful state, better organized and better armed. If there were similarities between 1919 and 1989, they were in the languages of protest. Students on both occasions claimed for themselves the voice of righteousness, always a potent political resource in China. They demanded the right to influence the policies of their government. And they enjoyed the support of the people. In light of that popular sympathy, the police backed down. A new May Fourth Movement was in the making. The lead editorial in that day's edition of *News Herald* (*Xinwen daobao*), the underground newspaper that the students managed to publish that spring, claimed the mantle of 1919. "Every popular democratic movement in the history of the world has propelled society forward," it proclaimed with optimism. The students dared to hope that their movement not only would be ranked with the May 4th Movement of 1919 but might be added to the glorious list of other revolutionary anniversaries 1989 brought: the fortieth anniversary of the founding of the People's Republic of China (1949), and the two hundredth anniversary of the French Revolution (1789).

Some had another anniversary in mind as well. Several prominent intellectuals, notably astrophysicist Fang Lizhi, addressed letters to supreme leader Deng Xiaoping in January and February. They reminded him that 1989 was also the tenth anniversary of the dismantling of the

Democracy Wall. Activists at the Democracy Wall, hoardings at the Xidan intersection two kilometers west of Tiananmen Square, had called on Deng in the fall of 1978 to match his Four Modernizations in the economic sphere with what one of them, Wei Jingsheng, had turned the "fifth modernization," democracy. The movement became viable for a time because of struggles within the Party leadership between Deng and pro-Maoist elements, but in the spring of 1979 Deng suppressed the wall and those who posted their views there. Wei was sentenced to fifteen years in prison and was still there as 1989 dawned. Fang Lizhi and others called for his release.

At the time of the Democracy Wall, Deng Xiaoping had only recently returned to power after two bouts of political disgrace at the hands of Chairman Mao Zedong. But Mao was safely dead, and Deng could tear up the Maoist blueprints for permanent revolution and set China on a course of modern economic development. His reforms changed China. The Communist Party's influence on decision making at lower levels was trimmed back; private markets were reintroduced; communalized land was broken up and returned to family management; private enterprise invaded industry; the military was reduced and professionalized. At the same time, China opened more widely to the outside world. Some, including the Democacy Wall activists, thought Chinese society might begin to move away from Mao's dictum that politics be "in command." They looked to separate state and society, if only in limited ways.

The future toward which Deng seemed to be steering China was mined with contradictions. By replacing the revolutionary zeal of Mao Zedong's peasant Communism with the technocratic mechanisms of his own, Deng robbed the Party of its reason to exist. A state-supervised market economy requires neither Communist ideology to guide it forward nor a revolutionary cadre to ride at the vanguard of change. It requires only a managerial elite to oversee economic decisions during the transition from state monopoly to state capitalism. Intellectuals began to see little reason to continue venerating the Communist Party for its past accomplishments. The Party had fought a brutal civil war with the Nationalists, risen from near-disaster during the historic Long March in 1935, then resisted the Japanese invaders from caves in the dry hills of north China. These were grand accomplishments, but they lived on only in history books and the minds of old Party leaders. Technocrats could better administer the economy than Party stalwarts. Legal specialists could better oversee judicial reform than political appointees. Artists and writers felt they had a larger claim on creativity than Party hacks.

As the decade of the 1980s wound to a close and Deng turned eighty-four, university students found themselves consumed with passions other than revolutionary fervor. The ideology of Communism had become meaningless. As far as they were concerned, the Party in the 1980s was left with no role except a parasitic one: to protect its members' privileges. More and more one heard complaints about corruption among officials, which exposed the absence of legitimately useful functions for the Party. The Communist Party had been the revolutionary vanguard. Now it was a vanguard without a revolution. With no role for Communism, the role of the Party evaporates.

Young China thus turned away from allegiance to the Party as Deng Xiaoping's reform program took hold. No longer required or motivated to espouse transcendent love for an ideology or a leader, young men and women looked elsewhere to compensate for the poverty of ideas and values that overtook them in the post-Mao secular age. Chinese culture had nothing to offer. These students, born in the high days of the Cultural Revolution, had been given a relationship with their own history that was at best ambiguous about the values of traditional Chinese culture, at worst contemptuous of those values. The Party pictured to them a past that was nothing but a prolonged exercise of feudal exploitation robed in a vacuous morality of Confucian subservience. They were bereft of an authentic tradition for themselves. China offered them nothing in which they could believe. They turned instead to the West: its consumerist culture, its heterogeneous values, its doctrines of democracy and human rights, even its religions. What they found abroad only further eroded their faith in what they had at home. Xenophobic campaigns against "spiritual pollution" in 1983 and "bourgeois liberalization" in 1985 could not discourage young China's infatuation with the West.

Economic growth intensified this infatuation as urban Chinese aspired to Western affluence. And despite political campaigns to the contrary, that infatuation was made worse by messages within the official culture. One had only to turn on Chinese television, as I did in Shanghai four months after the Massacre, to see in the advertisements for refrigerators and perfumes the persistent fostering of Western-style consumerism. The government told its citizens to aspire to revolutionary ideals; at the same time, state television trained them in bourgeois values. (Equally disturbing were the shows themselves. In one evening variety show performed by the Army Song and Dance Troupe, female dancers in wedding-dress costumes pantomimed great moments in Party history.) But the infatuation with the West in the 1980s was not purely consumerist. The educated came to yearn for some of the noneconomic benefits that affluence seemed to have bequeathed to the West. They wanted freedom from the Party's paternalistic control: managers wanted Party secretaries out of their offices, private businessmen wanted government officials out of their pockets, and teachers wanted Communist ideology out of their textbooks.

Deng Xiaoping is a pragmatic administrator rather than an inspired thinker, a deft politician rather than a Great Helmsman. But he is also a confirmed Leninist, committed to the concept of a vanguard Communist Party leading the nation toward socialism rather than mediating among competing interest groups. If he and Chairman Mao share anything, it is the presumption that policies can be assured only by monopolizing power at the top. Like Mao, Deng has ended up confusing his personal dominance of the Chinese state with the integrity of the nation itself. Most of the burgeoning political demands of the 1980s—demands his own reforms had fueled—challenged the Party's monopoly on power, and his own. They were unacceptable.

Frustrations over the limits of reform and the spiraling cost of living under Deng's economic reforms were widespread as 1989 dawned. By April, with the seventieth anniversary of May 4th approaching, students sensed that something had to happen. It was not an abstract idea about freedom or inflation that set things going, however, but a death. Hu Yaobang died.

Hu Yaobang died of a heart attack on the morning of Saturday, April 15, at the age of seventy-three—young by the standards of China's current leadership. He and Zhao Ziyang were the two men Deng Xiaoping had catapaulted into the top leadership to steer China on its course of reform in the 1980s. Hu had served as secretary-general of the Communist Party until January 1987, when he was unseated for failing to come down hard on student protesters. He was replaced by Deng's other protégé, Zhao Ziyang, whom Deng moved over from the state premiership to take charge of the Party. The official account of Hu's death observed that he had fallen ill a week before in a meeting of the Political Bureau of the Party Central Committee. Popular lore claimed that Hu had suffered his heart attack when Bo Yibo, a conservative opponent, asserted that the reform program Hu and Zhao had championed was "the error of the decade."

Hu might seem an odd catalyst for a democratic movement. He was not an advocate of democratic rights, but he was a major architect of China's reforms in the 1980s and had won the support of intellectuals by representing their interests and restoring reputations that had been trashed during the Cultural Revolution. He was held in respect for having sacrificed his position as Party secretary-general on behalf of protesting students in 1987. He had taken a fall for them.

Still, the significant thing about his death was not so much that Hu Yaobang died but that he died in the right place at the right time. In an environment in which politics presents a bland face and operates in the public sphere only by ritual and routine, the unexpected death of a political leader can split the facade and allow the routine to be challenged. In addition, the opportunity to mourn publicly provides a focus for expressing pent-up dissatisfaction. Hu's death did just that. As Fang Lizhi told me later, "Hu Yaobang himself wasn't that important, and the regard heaped on him was excessive. But in China, a leader's death serves as an excuse for people to assemble. The Party can't very well tell the people not to mourn a Party leader! Since a funeral is the only situation when people can assemble, you take advantage of the opportunity. It's only when people assemble that something can be achieved."

Naively, unrealistically, students associated Hu with their own in-
choate and unexpressed aspirations for a more open, pluralistic society.
"Young people couldn't help themselves," one participant recalled. "They
wanted to say something or do something in memory of Secretary-General
Hu. Two years before there had been student demonstrations in Beijing
and Hu was criticized by the government and quit his position." Memories
of those brief demonstrations in the winter of 1986–1987 were not what the
government wanted in the forefront of student minds that spring. The
official report on the news at dinnertime made no mention of his ousting. It
was as though, magically, Hu had never fallen afoul of the senior Party
leadership.

The news spread rapidly across university campuses as soon as it was
made public Saturday afternoon. Posters appeared bearing such extravagant headlines as "A Great Loss for Democracy and Freedom" and "The
Star of Hope Has Fallen." In its report on reactions on the evening after
Hu's death, the Japanese Kyodo News Service predicted that Hu's death
might revive calls for greater freedom and democracy, for Hu had fallen
from favor in the wake of student protests voicing such demands. But if
Hu's death provided an opening in a political system where such openings
are rare, it was because the men at the top of the system were divided
among themselves.

Sunday, April 16, saw the first organized march of student mourners
to Tiananmen Square. Another followed on Monday, drawing four thousand people. "Long live freedom!" they chanted. "Long live democracy!
Down with corruption! Down with bureaucracy!" The Public Security
Bureau (PSB), China's unarmed police force, appeared at the Square but
made no move to interfere.

Tiananmen Square was the logical place for the students to gather.
First of all, it is central. It stands directly south of Tiananmen (Gate of
Heavenly Peace) which fronts the Forbidden City, the imperial palace in
the heart of the old capital. Beijing's main east-west thoroughfare,
Changan Boulevard, crosses it. Most campuses, on the other hand, were
flung to the far suburbs, a few beyond the 2nd Ring Road, most beyond
the 3rd, and some even outside the 4th, which is under construction (see
Map 1). The students had to make a long trek to get to the Square (Beijing
University is fifteen kilometers from Tiananmen), but the effort was worth
it. The Square is not simply centrally located; it is the ritual center not just
of Beijing but of all China, and commands symbolic force that no movement could do without.

The government built Tiananmen Square in imitation of Moscow's
Red Square in the 1950s to mount displays of obedience to Party authority. During the Cultural Revolution it had been a parade ground for Red
Guards, but in April 1976, the citizens of Beijing showed that not just the
government could draw upon the Square's power. That spring, Tiananmen briefly was the site of spontaneous memorial observances for former

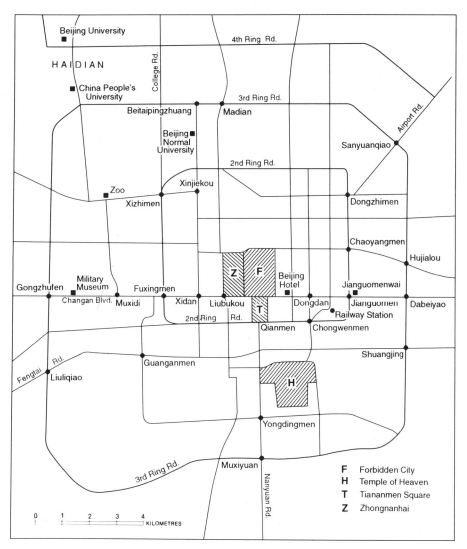

Map 1. Beijing.

Premier Zhou Enlai, which led to demonstrations in support of Deng Xiaoping and against the ultraleft element within the Party leadership (later excoriated as the "Gang of Four.") What was later called the April 5th Movement was suppressed by police and militia, and Deng was banished from power for the second time in his career. That protest may have been defeated, but the protesters had shown that the Square was holy ground not only for the government.

The small demonstrations on Sunday and Monday were followed on

Tuesday by a sit-in of several hundred students in front of the Great Hall of the People. On the west side of the Square, the Great Hall houses the National People's Congress (NPC) and the State Council, legislative and executive branches of the government. The demonstrators wanted to present the State Council with a seven-point petition anchoring a variety of proposals—curbs on corruption and officials' privileges, freedom of speech, improved funding for education—to the demand that Hu Yaobang's contribution to China be reassessed in a more positive light.

The students were not chased from the Square. Still, it took courage to participate in a march on Tiananmen and carry a sign calling for clean government. Party workers on campus kept lists of those who participated, and Public Security Bureau photographers took pictures, but the students felt safety in their growing numbers. Perhaps they were naive enough to think they could risk black marks on their permanent records. As the Chinese saying goes, only the newborn calf does not fear the tiger.

The word these bold students selected to express what they felt China needed was "democracy," although other concerns would get voiced during the spring of 1989. The Chinese term, *minzhu*, means "the people as masters." It came to China from the European political tradition via Sun Yatsen, founder of the first Chinese republic in 1911. Sun used it to express the consultative relationship that should exist between a people and its rulers, something unknown to imperial China. Democracy leapt to the forefront of Chinese political rhetoric in 1978 with the Democracy Wall Movement, and the government found it had to absorb the term into its own discourse, but to mean something desirable in the long run and not immediately feasible. In the first few days after Hu's death, little Democracy Walls appeared on campuses throughout China. At the time I was living at Nanjing University. There the official showcases lining the entrance to the South Campus, normally reserved for photo displays or earnest essays by model students, were covered with posters praising Hu Yaobang and lamenting the current leadership's failure to uphold fair and honest government. On the top of one of these cases, a student had pasted, in stylish calligraphy, a yellow sheet of paper inscribed with the Chinese characters *minzhu qiang*, Democracy Wall.

The activists of the 1978–1979 Democracy Wall Movement had formulated less a critique of communist politics than the simple request that individuals be permitted agnosticism on the question of whether communism was the best form of government. The poster writers did not call for Communist Party rule to be replaced by an American-style democracy; they asked only for public accountability. Their counterparts ten years later adopted largely the same outlook, though inflation and growing corruption in 1989 gave the call for political openness greater appeal. "This was not a great movement for Western-style democracy," as one American scholar in Beijing stressed, contrary to the impression that the international media would soon create. "The students made a big point of saying they were not against the country, or the Party. They were against corruption. They said they wanted simply a dialogue, a free press, freedom of speech, and an end to corruption. We did not hear any elaborate demands for institutional overhaul or democratization."

The police at the demonstration on Monday, April 17, had been unarmed. On Tuesday, Public Security officers were joined by the People's Armed Police (PAP). Unlike the PSB, the PAP is entitled to carry weapons. The PAP was formed in 1983 when over half a million troops were reassigned from the regular army as part of a larger plan to reduce the size of the military. (Perhaps not coincidentally, 1983 was a law-and-order year, a year of widespread and much-advertised public executions of at least five thousand people.) By designating them PAP rather than PSB, the government was able to expand its police force without increasing the power of the Ministry of Public Security, which controls the PSB and has been ill-disposed toward reform.

No conflicts broke out between the students and the PAP on the Square on Tuesday. Three NPC delegates came out to receive the students' petitions about 8:00 P.M., but the students did not go back to their campuses. At 10:50, about two thousand left the Square and walked west along Changan Boulevard to Zhongnanhai (Middle and South Lakes), the imperial pleasure park west of the Forbidden City where top government leaders live and work. They gathered in front of Xinhuamen (Gate of New China), the ornate entrance at the south end of Zhongnanhai, and asked to speak directly to Premier Li Peng.

Some students made the first of four attempts to push their way through the gate at 12:20 A.M. The guards in front of the gate locked arms and formed a human wall to prevent their entering. In scuffles over the next four hours, both police and demonstrators were injured. At 4:20 A.M., the municipal government broadcast a warning that the demonstrators should disperse. Some two thousand unarmed PSB officers arrived to enforce the order. The police lined up shoulder to shoulder and squeezed the students away from the Party headquarters. Aside from a few minor cuts and bruises, the incident was nonviolent. The demonstrators were gone by 5:00 A.M. Wednesday morning.

Within hours, another group of three thousand students marched into Tiananmen Square, again to conduct services in honor of Hu Yaobang. By early evening, the students had drawn a crowd of over a hundred thousand. Later, several hundred students, followed by about eight thousand onlookers, proceeded once again to the Party headquarters at Zhongnanhai. They reached the entrance gate of Xinhuamen at 11:30 P.M. Guards again prevented them from entering. The standoff continued until 3:40 A.M., when the government broadcast another warning to disperse. Some were persuaded to leave an hour later, but a remaining two hundred refused to go. A squad of unarmed PAP came in to remove them. The students were hustled roughly onto a bus and driven back to their campuses. A few were kicked and beaten.

New China News Agency, China's official news organ, remained silent for a week before releasing its account of the Thursday morning fracas at Xinhuamen. In its report published on April 27, the agency conceded that students and police in the early hours of April 20 had ended up "shoving and beating one another," but insisted that "rumors that police held batons and wore big boots are groundless. They were ordered not to carry their batons when on guard duty and all of them had on sneakers. None of the police on duty wore big boots. None of the students was seriously injured."

This report became a point of contention between the students and government in the weeks that followed. At televised talks with student representatives on the afternoon of April 29, the agency interpretation was challenged. A student from the University of Politics and Law insisted that the policemen had been wearing not only leather boots but leather belts as well, which they had used to beat the students. The student also asserted that after the crowd had dispersed from Xinhuamen, a fellow student was beaten by the PAP at the south end of the Great Hall of the People as he was walking down to the subway station. "He was beaten until his head was cut and blood gushed out. According to what he said, he was beaten by armed policemen with a leather belt. He has kept his bloodstained clothes [as] material evidence." The government's denial that leather boots and belts had been used was satirized in the demonstration on May 4 by a student who attached a pair of leather shoes, a leather belt, and green Army sneakers to a placard, arranging them in a mathematical formula: leather shoes + leather belt = sneakers. A student leaflet also challenged New China's account by printing a photograph showing seven policemen beating a student on the morning of April 20, with the caption "No-one lifted a hand!!!"

One other complaint the students made about the Thursday morning scuffle was that the demonstrators had been forced into a bus that was not nearly large enough to hold them all. A member of the conservative Beijing Municipal Party Committee was defensive: "We miscalculated the number of students. If we hadn't, we would have sent for a much bigger vehicle." To this he added, "We didn't know what kind of vehicle the Hongqiao Bus Company would send." This curious admission suggests that the government's response to the students was still very much ad hoc and without an overall plan. The government seemed to have little sense of what it was up against, or how it should handle it.

The municipal government declared on Thursday that the Square was to be closed to the public from 8:00 A.M. Friday, April 21, until noon Saturday. This was to accommodate the funeral service for Hu Yaobang, which was to be conducted inside the Great Hall on Saturday morning.

The statement, carried on the front page of the *Beijing Daily*, was not enforced. Students gathered in the Square Friday, and an estimated forty thousand stayed through the cold night so as to be inside the police cordon Saturday morning to put their demands before the guests at the official funeral. The PAP were stationed at the bottom of the steps leading up to the Great Hall on Friday to keep the students back. Conflict did not break out, but there was tension in the air. That evening, some PAP took off their belts and were nervously playing with them, but did not use them.

The funeral organizers negotiated with the students at dawn on Saturday. It was agreed that the students could stay in the Square during the memorial service. Given that Hu was an embarrassment to the Party establishment, the occasion was an incongruous affair. The Party hoped to staunch criticism of its demotion of Hu in 1987 by giving him a funeral on a scale quite beyond the rites normally observed for the death of an ordinary member of the Central Committee. It was a large concession to the students, but it failed to stop the outpouring of discontent. Too much had happened in the week between Hu's death and his consecration to let this honor stand as sufficient redress. So while four thousand government leaders were inside the Great Hall performing their obsequies on Saturday morning, tens of thousands of students congregated outside in the Square, seeking to press a petition for reform into the hands of government leaders.

Fifteen hundred policemen staked out a line between the students and the Great Hall. Thousands more stood shoulder to shoulder around the perimeter of the Square to close it off. Some of these young men were first-year students from the Beijing Police Academy, completely untrained in riot control and personally sympathetic with the Movement.

Other uniformed figures took up positions in the Great Hall. Some were carrying unloaded guns. This was the first appearance of the People's Liberation Army since the Movement started a week before. A report from Hong Kong said that the Beijing Garrison, as well as other Army commands within the Beijing Military Region, had been put on full alert as early as the previous Tuesday, and had been moved into positions of readiness around the city the night before Hu Yaobang's funeral at Deng's personal request. But the soldiers at the Great Hall were the first sighting. They belonged to the Beijing Garrison, a special corps charged with the defense of the capital. Consisting of two divisions (totaling roughly twenty-five thousand soldiers), the Beijing Garrison is the main military unit within the Beijing Military District. Although reduced from twice that strength in 1986, the Beijing Garrison is still politically powerful. Zhou Yibing, its chief of staff from 1981 to 1987, is now commander of the entire Beijing Military Region. (He succeeded Qin Jiwei, who was promoted to minister of defense.)

Students came away from this first distanced encounter with the People's Liberation Army believing that they could win the hearts of Beijing soldiers to their cause. The soldiers' non-belligerence seemed to confirm

their suppositions about the changing nature of the Army under the reforms. Some students at Beijing University (known by its shortened title
as Beida) had formed a discussion group that called itself the Scientific
Seminar that winter. They had already undertaken an analysis of the
Army before April 15. As Beida student leader Shen Tong recalled, the
group was optimistic. "We saw that new recruits were mainly peasants
who had benefited greatly from the reforms. Having a personal investment
in the new economy, they were unlikely to be brainwashed easily or to
obey orders blindly. Another part of the Army was composed of young
officers who had recently graduated from military schools. They were well
educated and likely to be sympathetic to the Democracy Movement." The
students were aware that the Army could pose a threat to the Movement,
"but we felt that if we did not go too far, the Army would not present a real
danger."

Some Beijing residents feared otherwise. A Canadian professor who
arrived in Beijing on the day of Hu Yaobang's funeral was informed by her
Chinese host that Beijing was "under martial law." The term he used was
technically incorrect, of course. Martial law had not yet been imposed.
Perhaps her translator simply did not have any other vocabulary to cover
the situation. But his choice of phrase indicates that there was already a
widespread sense that the military might be used to end the disturbances.

Word began to circulate the following week that soldiers were coming
into the capital region. On its noon broadcast on Tuesday, April 25, Hong
Kong Commercial Radio stated that at least ten thousand troops from
outlying areas had been ordered into the Beijing Military District. More
reports the next day insisted that troop movements were under way. The
Hong Kong wire service AFP on April 26 relayed that it had received
reports from several sources that twenty thousand troops from the 38th
Group Army had been moved into the Beijing Military District over the
previous two days. Elements of the 38th had already been in the district
for several days, probably even before Hu Yaobang's funeral.

The 38th Group Army would be a key player in the events to come. It
is one of six group armies stationed within the larger Beijing Military
Region, which covers the provinces of Hebei, Shanxi, and Inner Mongolia. It is one of the group armies that were formed in the wake of a major
reorganization of the PLA in the mid-1980s. A group army is supposed to
have between fifty and fifty-six thousand troops, though most are at 70 to
80 percent strength. The group armies operate under a highly centralized
command structure. They can move only by order of the chairman of the
Central Military Commission (CMC), a position that, at the time, was
Deng Xiaoping's sole formal post. According to a former soldier in the
Beijing Military Region we interviewed, no unit larger than a company
(about 120 soldiers) can be moved by order of anyone other than the CMC
chairman. Moving a company requires the approval of the commander of
the military region; and moving a squad (a dozen men) needs the okay of

the provincial commander. This system blocks independent maneuvering or combining of troops at lower levels. Everything is tightly controlled.

The 38th Group Army was formally established in August 1986. It was the PLA's first mechanized combined arms group. In addition to its four regular divisions, it includes a number of specialized branches, such as an antitank corps and an engineering corps. Considered the most modernized and best equipped of PLA units, the 38th was regarded as the Party's "model army." It is based in Baoding, 120 kilometers south of Beijing. This places it outside the Beijing Military District surrounding the capital, but closer to Beijing than any of the other five group armies in the larger Beijing Military Region. Whether rightly or not, the 38th is generally regarded by Beijing people as "their" army. Some of its soldiers and officers, including its field commander, General Xu Jingxian, had friends and family among the protesters. General Xu had broken his leg during training exercises at the end of March and was in the Army General Hospital in western Beijing when the student movement began. Being in Beijing, he had direct access to information about the Movement, which may have influenced his response to martial law orders in May.

As elements of the Army moved into the Beijing area, students learned on Tuesday, April 25, that the government flatly refused to engage in dialogue with them. What they did not know is that Deng the same day put the PLA on alert. That evening, the national media broadcast the official judgment on the students' demonstrations, breaking the virtual silence that official media had maintained over the students' activities since Hu Yaobang's death. Printed as the lead editorial in next day's *People's Daily*, it became known, and vilified, as the "April 26th Editorial." The editorial expressed Deng Xiaoping's personal response to the student protest that he gave at a Politburo meeting on Sunday and again at a meeting with Premier Li Peng and President Yang Shangkun the following day. According to the stenographer's report of Monday's meeting circulated to high Party officials, Deng declared that he did not fear "spilling blood." Nor did he fear the international reaction that spilling blood might provoke.

The April 26th Editorial was not calculated to ease tension. It complained that "an extremely small number of people with ulterior motives" was "taking advantage of the young students' feelings of grief for Comrade Hu Yaobang to spread all kinds of rumors to poison and confuse people's minds. Blatantly violating the Constitution, they called for opposition to the leadership by the Communist Party and the socialist system." The editorial charged the students with cynical and self-serving duplicity. "Flaunting the banner of democracy, they undermined democracy and the legal system. Their purpose was to sow dissension among the people, plunge the whole country into chaos, and sabotage the political situation of stability and unity. This is a planned conspiracy," the editorial decreed, and people should "take a clear-cut stand to oppose the disturbance." The editorial also warned that "those who have deliberately fabricated rumors and framed others should be investigated to determine their criminal liabilities according to law." The Democracy Movement amounted to nothing more than "beating, smashing, burning, and looting." (This standard turn of phrase for denigrating autonomous political activity in China would reappear word for word in Martial Law Order No. 1.) The editorial ended with the warning that "our country will have no peaceful days if this disturbance is not resolutely checked."

The students were outraged by the government's refusal to regard any of their requests as legitimate. They vowed indeed to rob the days ahead of peace. While the students spent Wednesday planning a major march on Tiananmen Square for the following day, Party leaders held meetings. At the meeting in Beijing, ten thousand Party cadres gathered in the Great Hall of the People and were told that Army units were poised to move into Beijing to take over security tasks from the police if circumstances warranted. Deng Xiaoping did not attend this meeting, but he is reported to have said in a conference with Premier Li Peng earlier that day that the student unrest must be crushed "by any means." His views were apparently relayed to a Party Committee meeting at Beijing University.

Deng's comments were not simply those of a civil leader who wanted stability restored and might turn to the military to see that it was. They were the words of the chairman of the Party's Central Military Commission. Despite the old Party adage that the Party controls the gun, the relationship between political and military leadership in China has never been that unambiguous. It is largely true that the People's Liberation Army is not in a position to act as an independent force in Chinese politics, and no exclusively military leader has succeeded in pitting himself against the Party and taking over the government. But it is equally true that no Chinese political leader can act independently of the military. The fact of the matter is that the Party man who controls the Army controls the Party. Deng was enough of a political realist to recognize this when he reemerged into power after the death of Chairman Mao. In 1981 he gave his protégés, Hu Yaobang and Zhao Ziyang, the leadership of the Party (secretary-general) and state (premier), respectively. He took for himself the leadership of the military, for from that position he could control all other cogs in the clockwork of Party organization. Deng officially stepped down from the chairmanship of the CMC on November 9, 1989, though even thereafter he continued to direct its affairs from an advisory group behind the scenes.

Deng did not administer the military singlehandedly. Toward the end of the 1980s, he was passing more and more of the day-to-day business of the Central Military Commission to Party Secretary-General Zhao Ziyang, whom Deng appointed as a CMC vice-chairman in October 1987. (Zhao would be stripped of that post in June 1989.) Zhao Ziyang's much-publicized effort to modernize the economy was paralleled by his work in modernizing the Army. Modernization in this context meant reducing the Army's size, diminishing its nonmilitary functions, trimming its prerogatives, and downgrading its political commissariat to a subsidiary role. The process of creating a leaner professional force began with a decisive CMC meeting in June 1985 when the PLA was set on a new course . It signaled a big departure from the old Red Army style of peasant guerrillas, and succeeded in freeing up close to 10 percent of the national budget. Deng supported Zhao's military reduction program, which was in midcourse in the spring of 1989 and breeding annoyance among older and more conservative elements in the PLA.

Zhao Ziyang's personal influence over the military was weak, however. The Army listened to him only because he was Deng's representative. At this particular moment in late April, however, Zhao was not even in China. He had departed on a state visit to North Korea after the Politburo meeting on Sunday at which the April 26th Editorial was first discussed, and would not return from Pyongyang until the following Saturday. His absence was crucial. It was known that Secretary-General Zhao Ziyang and Premier Li Peng had been on opposite sides of high-level policy conflicts for years. Zhao was a strong advocate of systemic reform; Li counseled go-slow moderation. With the Chinese economy in near chaos and society bubbling with discontent by the spring of 1989, Zhao's position within the leadership had been fragile for some time. Li had been on the offensive since the previous autumn. If Zhao chose to counsel lenience with regard to the student demonstrations, it was not because he entirely supported their ideas; in fact, he was keen on neo-authoritarianism (the notion that market economic reform can succeed only by concentrating power in a central authority) and is credited with introducing the concept to Deng. Rather, Zhao feared that Li Peng's hand would be strengthened within the Party if Li were permitted to let that hand come down hard on the students.

It had never been the students' intention to do so, but their activities intensified the factional tension between these two men. Time and again the students insisted that what they were struggling for transcended factional politics and, more critically for their own defense, was not tied to any existing faction. But their Movement raised for government leaders the vexing question of how their criticisms should be handled. In doing so, it cast a harsh glare on the larger and more sensitive problem of how, or even whether, political reform should go ahead. Li's faction had grown in power at the expense of Zhao's with the slowing down of economic reforms in the fall of 1988. In calling for moderation in dealing with the students, Zhao Ziyang hoped to use the Movement as a political lever against the Li Peng faction. Others, particularly Li Peng's supporters on the powerful Beijing Municipal Party Committee, saw the Movement as an opportunity to counterattack.

They would succeed. During Zhao's sojourn in North Korea, the

Politburo Standing Committee called an emergency meeting on Monday and set up an ad hoc group to deal with the crisis. This meeting was dominated by Li Peng and Beijing Mayor Chen Xitong and cut out entirely Zhao's reform faction. Although the meeting had constitutional legality, it was an unusual step to take. The decision to declare martial law originated within this Politburo subcommittee, possibly within the first few days of its being formed.

For the moment, however, the students held the initiative. The demonstration on Thursday, April 27, to protest the April 26th Editorial was everything the students hoped it would be. Over a hundred thousand students from most of the forty-odd institutes of higher education where classes were being boycotted responded to the directive of the fledgling Beijing Federation of Autonomous Student Unions. The government did not make any serious attempt to staunch the flow of marchers out of the northwestern suburb of Haidian. A thin line of policemen strung across the road south of the Beida campus could not halt the mass of students emerging from Beida at 8:30 A.M. A combined column of Beida and Renda students was able to push through a second blockade of some two hundred policemen south of Renda about 10:00 A.M. As onlookers cried "Clear off!" and "Get lost!" the police had no choice but to part and let thousands press through on their way to the Square. A dozen more police lines gave way as the students marched to the Square. It was clear that the police were going to let this demonstration go ahead.

Students from the four colleges in the northeastern suburb came to the same conclusion. The Beijing College of Chinese Medicine had been praised the day before in the *People's Daily* as the only college in Beijing where classes were still being held. This distinction so annoyed the students that they climbed over the school walls (the school authorities had earned the distinction by locking the gates) and marched to the Square. Again, they were able to push past a police line three rows deep at the 3rd Ring Road. They felt invulnerable. According to a student at the college, "It was widely rumored that the head of the PSB refused to suppress the April 27 march with force."

A little pushing succeeded in overwhelming irresolute police opposition. Still, the students knew that they had to maintain nonviolence in order to preserve the moral edge they enjoyed. They needed to show that they were not a conspiracy to overthrow the socialist order, as the April 26th Editorial characterized them. They opposed only the corruption and privilege that had crept into the socialist order. Accordingly, students carried signs announcing their support for the Communist Party. As Berkeley historian Frederic Wakeman observed, the new tactic to demonstrate in support of socialism and the rule of the Communist Party was "a brilliant decision. It not only lowered the stakes by substituting an evaluation of this particular government's performance for a normative attack upon the Communist Party and socialist system as such; the new slogans also guaranteed the support of a wide segment of the urban population suffering from rampant inflation, bribe-taking, and barely concealed economic exploitation." The working people of Beijing came out as sympathetic onlookers for the same reasons. Estimates of the numbers who lined the marchers' route through the city range from a third to half a million. The headline that day in the *Beijing Daily*, "The Working Class in the Capital Firmly Opposes Social Disturbances," had nothing to do with what Beijing residents actually did that day.

When the various streams of students converged at the Square, they found fifteen trucks and a tight cordon of hundreds of men in uniform awaiting their arrival. Rather than try to occupy the Square by force, the students simply walked by it along Changan. The march-past of 150,000 people took more than two hours. The procession stopped at the Jianguomen overpass on the east side of the downtown and turned north on the 2nd Ring Road to head back to the campuses. At 6:00 P.M., just as the last of the parade was turning north, the Army appeared.

A foreign diplomat who watched the scene from his apartment in the Jianguomenwai diplomatic compound that overlooks this overpass describes the encounter: "The procession had already turned up the 2nd Ring Road and the crowds of onlookers at the Jianguomen intersection were filling in behind them when a convoy of fourteen Army trucks came west along Changan Boulevard. It stopped for a while away from the intersection. As the last of the procession turned north up the 2nd Ring Road, the convoy tried to pass through the intersection and turn left to go south on the 2nd Ring Road. The people in the intersection—they weren't students—surrounded the convoy and stopped it for well over half an hour. They blocked the first couple of trucks, climbed on their hoods, and started making speeches. Citizens were also having long talks with the soldiers in the backs of the trucks, two at a time. After five or ten minutes, they let them go through, then the next two came up and they continued the process." The diplomat was struck by the fact that "when they let a truck through, it just continued on its way and didn't wait to make sure its compatriots got through."

The meeting between soldiers and Beijing residents was never intended to occur. One eyewitness noted that "the soldiers looked disoriented. They stared around as though they had never seen skyscrapers before, as though they had never been to Beijing." The soldiers belonged to the 38th Group Army but were not part of the units of the 38th normally deployed within the Beijing Military District. These troops belonged 120 kilometers south of the city and were out of garrison. Foreign military observers at first assumed from this sighting that the 38th was holding Army-level exercises south of Beijing and that this convoy was part of those exercises. It appears, however, that these elements of the 38th were moved into the Beijing area not for exercises but because of heightened concerns about security in the capital. What is surprising is that this convoy's command did not realize that its route would be inundated by thousands of demonstrators; surprising also is that rather than pushing through the crowd, the convoy did not immediately withdraw.

The Army did not appear to the student marchers again that day. It would not appear again in Beijing for another three weeks. But it was not dormant. By the beginning of May, anyone with a family member in the armed forces knew that all leaves for military personnel were canceled. The Army was on standby notice.

The New China News Agency account of the April 27 march announced that "no clashes were reported between the police and the students." This report appears to be untrue. A foreign medical student saw evidence to the contrary while on duty in the intensive care unit of a hospital (to be left unnamed) in the northern half of the city. Three or four students were brought into intensive care that evening, badly beaten, one allegedly with a broken back. "Friends of one of the students in intensive care came and asked for that student but were told by the hospital staff that they didn't know anything. The hospital staff had been informed by the chief physician that there was an order out to all hospitals in Beijing that, should a university student be treated for anything serious at all, nobody could say anything to anybody about it, not even to the patient's family." Shortly after, the students were moved to the military hospital. There was a report in the *Hongkong Standard* that four students dispatched to solicit support in the provinces went missing about this time.

As early as the end of April, then, violence was already casting a selective and secretive shadow on the Democracy Movement. The government was unwilling to engage in public gestures of violence, but possibly it was already condoning private acts to weaken student opposition. No one could predict what was to come. One foreign scholar's colleague told her the day after the *Hongkong Standard* report of student disappearances that "blood will be shed," either by the police or by the PLA. Where this would happen, or how soon it might come, he could not say, but he felt that it was coming.

Against this unsettled background, the fateful day of May 4 approached. Here was the students' best occasion to voice demands for democratic reform. At the top of their list now was official recognition of their autonomous organizations over the Party-controlled student unions already in place. Many realized that the odds of winning the battle were against them, as a Beida student remarked to me many months later, but the first successes had been too great to turn back now. "We were nearing the seventieth anniversary of May 4th," he said. "We all knew from past experience that every one of the many student movements in modern China's history had ended in failure. Our only hope was that each new movement might help ever so slightly to change China's fate."

The older generation was less optimistic. A Chinese scholar sat down early on the morning of May 4 to write a letter to a friend abroad. "I have seen the students on the street coming back from Tiananmen," he wrote, recalling the April 27 demonstration. "Most people in Beijing support them—including the intellengtsia, workers, and housewives. The policemen and soldiers do as well, though they have to obey the orders of the government. However, I know that the Movement can't do very much, because the Party's control of society is very, very strong. For example, there is an order nowadays that says that parents have to keep their children at home, and that members of the Party can't support the students, and if they do, they will be removed from the Party." Sanctions such as these underly the Communist Party's attempt to reestablish its moral authority as the symbolic date of May 4 approached. Yet to snatch the initiative from the students, the Party had to do more than issue orders forbidding Party members from allowing their children to go into the streets. It had to engage in ideological counterattack and demonstrate to an increasingly antipathetic public that the legacy of May 4th belonged to the Communist Party, not the students. Secretary-General Zhao Ziyang was given the job of bringing the students to heel. He was made the keynote speaker at the May 4th commemorative ceremony in Beijing, held on May 3 in the Great Hall of the People.

Zhao gave a fence-mending speech in which he sought to woo the students to the government's point of view, without granting them concessions that might irritate his more conservative colleagues in the Politburo. His plea was for the old Party virtues of stability and unity. Democratic reform was a high priority for the Party, he insisted, but the struggle to achieve full democracy had to be gradual. The lesson he drew from May 4, 1919, toed the conventional Party line: that only the Communist Party could lead China to the goal of genuine people's democracy. Zhao made no reference to the recent student movement, nor to Hu Yaobang. But his upbeat tone was at least a welcome relief after the previous week of censorious commentary in the official press. This was a speech intended as much for the Party hierarchy as for the students, to convince them that a middle course between suppression and capitulation was possible. The latter, unfortunately, had no sense of how precarious Zhao's position within the leadership was becoming. They did not act as he hoped they would. They decided that their march for May 4th would go ahead as planned.

Thursday, May 4, largely repeated the story of the previous Thursday. This time there was less anxiety about what the police might do. The students dared to believe that security forces would neither resist nor retaliate, and they were right. Tens of thousands of marchers, applauded and followed by larger numbers of citizens, poured down the streets from the northern suburbs, broke through thin police lines, and marched down to Tiananmen Square to wrest the heritage of May 4th from the Party. "There was no show of force by the Army or attempt by the police to keep spectators out of Tiananmen Square," reported Fred Shapiro, a foreign resident in Beijing. The only moment of potential conflict came as the demonstrators approached the entrance to the Great Hall, "which was guarded only by a few files of police—certainly a force no larger than the ones that the marchers had by now become accustomed to sweeping aside." The standard-bearer leading the demonstration paused dramatically in front of the entrance. Had he turned toward the Great Hall, "I don't believe that anything less than weapons fire—and mass casualties— could have kept the demonstrators out. For fifteen minutes, while the standard-bearer posed there, making the most of his moment, I doubt if Zhao or any of the hundred thousand of us in the Square or the Great Hall knew for certain which way the march would go. In the end, with a circular flourish, the bearer pointed the demonstration flag away from the Hall, and the parade followed it unopposed into Tiananmen Square." Elsewhere around the country, similar demonstrations took place on May 4, with similar outcomes. An observer in the northeastern city of Changchun, where thirty-five thousand rallied that day, noted "absolutely no hostility from the military or the police."

When the students got back to their campuses that evening, they listened to a radio broadcast of another speech Zhao Ziyang had given just that morning, this time at a meeting of the Asian Development Bank in the Great Hall while the students were outside in the Square. In this speech, Zhao took a significant step forward from the position he had espoused the night before in his official May 4th address. He indicated that he did not view the demonstrations as a serious political instability, and he gently advocated consultation with the students. Their desire to correct errors in the government, he suggested, matched the Party's own commitment to serve the people better.

Zhao's speech, combined with a growing sense among students that there was little more to be gained by continuing their protests, encouraged some to give up their boycott of classes. Foul weather also came onto the government's side. But a radical contingent within the Movement leadership was not willing to find common ground with the government so easily. They had to recapture the initiative, and saw a unique and unplanned opportunity approaching: the state visit of Soviet leader Mikhail Gorbachev the following Monday. The students respected Gorbachev for the political reforms he had initiated in the Soviet Union. While Deng and his appointees spent the 1980s concentrating on liberalizing the economy, they had left the Party-dominated political system little altered. Gorbachev on the other hand had kept the pace of economic reform in the Soviet Union slow, but he had allowed considerable political liberalization to take place. His *glasnost* is what the students wanted for China.

It was a prime political opportunity. The students knew that the Party attached great importance to Gorbachev's visit. This would formally mark the end of the Sino-Soviet rift dating back to 1960, when Mao Zedong and Nikita Khrushchev parted company under a cloud of mutual distrust.

Deng had finally achieved a historic reconciliation: Gorbachev's visit was to be the jewel in his crown. Interfering with that visit could be used to pressure the Party. The students were also confident that the government could not afford to indulge in a bloody repression with Gorbachev on his way. At a press conference on Friday, May 12, State Council spokesman Yuan Mu expressed his hope that the students would "safeguard China's international image" and call off the demonstration, but he also indicated that the government would continue to exercise restraint. This was a green light for the leaders of the Movement.

To recharge popular support, the students drew from the world history of nonviolent resistance a new tactic, the hunger strike. "The idea of a hunger strike was really something," a Beida student recalled. "I had read about hunger strikes in South Africa in my history books. I could not understand why the students were going to do that. I thought it was an impossible idea." This impossible idea caught on quickly. "A graduate student from Beijing University said we should start a hunger strike because we don't want this cause to be destroyed without achieving any result. Students started talking about it, and that afternoon, or maybe the next day, I saw a big list. It started with over one hundred names, and later two hundred, and then more." (Beishida students insist that the idea started with them. Not to be outdone, Beida students early the following week proposed self-immolation if their demands for recognition and dialogue were not met.)

The hunger strike was not a tactic Chinese students had ever used to advance their causes, but it became permanently incorporated into the Chinese repertoire of protest on the weekend of May 13–14. A foreign observer arrived at the Martyrs' Monument at 6:30 P.M. Saturday evening and watched the first groups of hunger strikers enter the Square. "Those who were fasting sat themselves down in the central area, which was protected, and everyone else milled around them. The students came in on their bikes in a well-organized fashion, carrying their banners. The mood was festive." Among the small foreign contingent on the steps of the Monument was a man who belted out a Polish labor song, to the cheers of the students who had assembled there. This was their stab at a Chinese version of Solidarity, a specter the Chinese Communist Party could only dread.

The hunger strike was a brilliant, if dangerous, tactic. Refusing drink as well as food, hundreds collapsed daily. According to official estimates, between May 13 and May 24, thirty-two hospitals in Beijing treated 9,158 cases of collapse, of which 8,205 required hospitalization. Ordinary people could not help but become emotionally involved in what the students were doing. They poured into the Square in large numbers. By radicalizing the protest, however, the strike closed off the possibility of compromise, for either side. It left the Communist Party no room to do anything but capitulate—a position that no ruling clique likes to be in. There was a brief round of negotiations on Sunday, the first day of the hunger strike, but the talks broke down quickly when it became clear that the government would not give in on the students' first condition for further dialogue: the rescinding of the April 26th Editorial. The students were more firmly resolved than ever to remain in the Square when the Soviet leader arrived.

The demonstrators' unwillingness to vacate the Square for Gorbachev's formal welcome ceremonies on Monday was a grand humiliation for Deng Xiaoping. The international media attention they stole from what was to have been a historic event intensified that humiliation. It was also a fateful decision for Zhao Ziyang's position within the Party. Unable to secure concessions on either side, Zhao found himself isolated within the leadership and unable to block the ineluctable slide toward martial law, a decision favored by most of his political rivals.

Police were out in force around the city Sunday night, but nothing happened. The demonstrators were left unmolested to defy the twelve-hour government order closing the Square as of 8:30 A.M. Monday morning. The PAP did nothing except protect the Great Hall of the People, where Gorbachev would be officially welcomed. PAP numbers increased later in the morning. Their only intervention was to turn back a crowd of several hundred that rushed up the steps of the Great Hall early that evening.

The Chinese government could do nothing but bow to the students' occupation of the downtown core and change its plans for Gorbachev's reception. Gorbachev was formally welcomed Monday on the airport tarmac instead of the Square, where the ceremony should have been conducted. Then his meeting with President Yang Shangkun in the Great Hall of the People was delayed for two hours because his cavalcade had to be detoured to avoid the demonstrations. His Tuesday afternoon meeting with Zhao Ziyang and Li Peng had to be switched from the Great Hall to the state residence where he was staying. And his visit to the Forbidden City on Wednesday had to be canceled when a record one million people showed up in the Square that afternoon. The Chinese government also had to cancel the ceremony in the Square planned for Tuesday, at which Gorbachev was to have mounted the steps of the Martyrs' Monument to place a wreath in honor of those who had died in the wars of liberation. The students debated among themselves whether to open a path for Gorbachev. The majority wanted to but deferred, as is done according to the prevailing Chinese concept of democracy, to the minority. The argument was probably moot anyway because Soviet security personnel feared that Gorbachev's safety could not be guaranteed under such conditions. For his hosts, these cancellations and changes of plan generated enormous embarrassment and confusion.

On the afternoon of Wednesday, May 17, the day Gorbachev's visit to the Forbidden City was canceled, the Democracy Movement achieved its greatest flowering. Spontaneously, a million people thronged to the

Square from every direction and by every available means of transportation. It was the hunger strike that did it, as one Beida student commented later. "The key was those seven days of hunger strike by several thousand students. It brought the people over to their side. Without the hunger strike, the rest would not have happened."

While Gorbachev was in China, the Chinese government could do nothing to bring the Movement to heel. Nor did the police interfere. In fact, they cooperated to some extent with the students. Qinghua University students sent a request to the municipal PSB on Wednesday night for help in maintaining order along the ambulance corridor reaching the hunger strikers in the Square. At 6:30 A.M. Thursday morning, they led in eight hundred policemen to assist the student pickets. Later that day, members of the Beijing PSB came out to march in support of the students, brandishing a banner that called for doing away with the practice of keeping police and Party files on individuals. It was rumored that the PSB nonetheless maintained plainclothes observers on the Square.

Students were encouraged to assume that the apparent sympathy of the police rank and file toward the Movement was shared by at least a segment of the Army. Some Army administrators even took part in the Wednesday parade of support. Staff from the PLA General Logistics Department, some dressed in camouflage, marched down Changan Boulevard in the grand procession chanting, "Down with corruption!" and "We demand democracy!"

Left untouched, the students were now coming to the dangerous conviction not just that the military would not be used but that it was on their side. On Monday, a student on the Square announced by loudspeaker that the 38th Army had resisted a government order calling it into Beijing, that "the officers and men, from the Army commander to rank-and-file soldiers, have all refused to enter the city." A seven-point bulletin posted around the city that evening repeated the allegation: "The 38th Army has refused to enter Tiananmen Square and carry out clearing operations." It also said that "a young Beijing policeman in a letter to the students says, no matter what bureaucrat orders us to suppress you, we will not act." Other letters allegedly written by soldiers of the 38th in support of the students were distributed on the Square as further testimony that the Army was with the students and would not suppress them.

According to student sources, the disenchantment within the 38th Group Army went right to the top. I cannot confirm the story that follows, but it is difficult to ignore, in part because it was so widely believed. The commander of the 38th, Xu Jingxian, son of legendary PLA general Xu Haidong, had broken his leg earlier that spring while conducting training exercises and was in the PLA General Hospital in western Beijing during April and May. He was still there in mid-May when he was called to the headquarters of the Beijing Military Region. There, on Thursday, May 18, representatives of the PLA General Staff and General Political Department presented him with an order, signed by Deng Xiaoping, to mobilize his army in preparation for entering Beijing and bringing the Movement to an end. With his leg still not fully healed, General Xu went back to his own headquarters to oversee the preparations and draw up a schedule of deployment. When this was completed, he telephoned PLA Headquarters and reported that because of his game leg he could not personally lead the 38th into Beijing. He was warned that refusal would be regarded as insubordination, but he stuck to his guns and said that he would refuse to take command of the operation regardless of what charges might be made against him. He hung up the phone, booked his sick leave, and readmitted himself to hospital. The 38th would not fight under his command.

It is difficult to gauge the extent of actual support within the military, but the signs continued to reassure the students that it was general. On Thursday morning at 6:00 A.M., the students' broadcasting station on the Square read what they said was an open letter that some PLA officers had sent to the Central Military Commission. The officers urged that the PLA not be used to suppress the students and that the CMC act in accordance with Zhao Ziyang's posture of restraint and accommodation. They called on the CMC to mobilize staff from Army hospitals to assist in aiding students on the Square. The officers also demanded that the military reforms of the previous years continue and that military spending be cut in favor of greater investment in education. The students could not have hoped for a stronger gesture of support, but it fed a dangerous presumption. The students now dared to think that the government could not mobilize the armed forces against them.

For the moment, there was little the Chinese government could do, especially when Gorbachev himself was speaking in guardedly favorable terms about the students' activities. Although Gorbachev did not accept the students' invitation to meet with them, he responded to questions about the demonstrations in a press conference Tuesday evening by stressing the need for socialist countries to continue in the necessary process of making the people masters of their own society. Democratization and openness favored that process. He argued that the demonstrations should be viewed not as a crisis but as a turning point. Deng Xiaoping was clearly of a different opinion; not surprisingly, for leaflets circulating among the demonstrators on Wednesday explicitly called for his resignation. But he had to wait. The delay only raised the stakes.

Premier Li Peng appeared on television on Wednesday and again on Thursday. His remarks during those appearances indicated that he was not budging on student demands, and suggested that he was gaining the upper hand in the factional struggle between himself and Zhao Ziyang, with Deng Xiaoping's backing. It had not been the students' goal to force a crisis in the factional struggle within the Party leadership; student leaders stressed this repeatedly throughout the Democracy Movement. But the split was occurring, as the students themselves began to realize. A Beida student recalls being alarmed by Li Peng's performance on television on Wednesday afternoon. "When we saw Li Peng talking in the Great Hall about protecting the achievements of socialism, we immediately went out in the streets. People had the feeling that Zhao Ziyang had lost."

Indeed, he had. At a meeting with Deng and other top leaders that day, Zhao had called for continuing toleration. He was voted down. The defeat was decisive. Deng apparently insisted that the Army now had to be used. "I have the Army behind me," he is reported to have boasted.

"But I have the people behind me," Zhao replied.

"Then you have nothing," retorted Deng.

Next morning, Gorbachev departed from Beijing. Torrential rains fell on the hunger strikers, who took shelter in buses provided by the Chinese Red Cross. In a television news feature that morning, Li Peng put on an unconvincing show of commending the fasting students for their patriotism. It occurred during a surprise midnight visit to hospitalized hunger strikers he made with Zhao Ziyang and other leaders. But his weak performance on television did not matter, Li Peng had already prevailed behind the scenes. With NPC Chairman Wan Li away on a state visit to Canada, there was no one in Beijing who outranked Li Peng constitutionally. Zhao could not block him.

Li Peng's second television appearance on Thursday evening conveyed the impression of a man who needed to make no compromises. It was a broadcast of a late afternoon meeting in the Great Hall between Li and twenty-two student leaders. What was to have been a dialogue turned into a confrontation that left neither side room for future maneuver. Li insisted that the students end the hunger strike and indicated that the government's patience was running out. His popularity was now at an all-time low, but he was prepared, politically and personally, to face negative consequences. These he anticipated. A former Army officer with connections inside Zhongnanhai informed a foreign friend that Li Peng's daughter, who normally lives outside Zhongnanhai, moved into her father's compound that day for her own security.

The rhetoric on the Square Thursday evening heightened as tensions rose. For the first time, workers came out in large numbers not just to share in the students' heroism but to speak on their behalf. I have listened to smuggled tapes of that night's speeches and the language is passionate. One worker called China "a lion just now waking up" and said that the hunger strikers were the voice of China, lending this lion a voice to roar in righteousness. "The people thoroughly understand that you are conducting your hunger strike not to oppose an honest government but to force it to become honest, not to create chaos but to bring freedom, democracy, and glory to our nation and our race." In the language of *The Communist Manifesto*, this people's orator declared that "in this movement to fight for democracy and freedom, all we proletarians have to lose is our chains, our slaves' shackles. We have the entire world to gain. Arise! Let us be slaves no longer!"

Another who addressed the crowd on the Square that evening raised the rhetorical temperature further by calling on the Communist Party to step down in the face of the students' protest. "The Communist Party, like a strong tree, was a good pillar, and people appreciated it, but once it started to rot, this great tree became infested with insects. It can no longer lead China. (Crowd: "Right!") This corrupt party can no longer lead China. (Crowd: "Right!") So that makes me an 'anti-Party element'! (Crowd: "She's a tough character!") Good people must stand up and declare that the Communist Party should step down." She returned to her opening image. "Like a great tree, the Communist Party when it was young had deep roots and luxuriant foliage and reached up to heaven. It was a good pillar. But truth is relative, not absolute. And the nature of a political party is the same. The Communist Party conquered the country for us and made great contributions. But the leaders are now completely corrupt. The roots of this great tree have rotted, and insects infest not just the roots but the tree itself. It should be removed." The crowd that had gathered around her broke into applause. "Everyone has been afraid to say that the Communist Party should step down. But me, I'm not afraid. Students are about to die, so what do I have to fear? (Applause) As a person of conscience, I do not fear death. (Applause) Suppose someone wanted to arrest me. Well, for the sake of the lives of the three thousand [fasting] students, let them arrest me. I'd go willingly. (Applause) I believe we are at a critical moment. We must oppose the view that only the Communist Party can lead China."

For the first time in forty years, people were being cheered for standing up in public and advocating an end to Communist rule in China. The Democracy Wall activists in 1979 had taken an agnostic view of the Party's monopoly of power. In 1989, some Democracy Movement supporters were rejecting it unconditionally.

By the early hours of Friday morning, the Rubicon of toleration was crossed. At 4:45 A.M., Zhao Ziyang and Li Peng made a surprise visit to the students while night still hung over the Square. The visit was clearly at Zhao's initiative. Li said practically nothing and just waited around while Zhao talked to the students. Presumably Li did not trust Zhao out of his sight and had consented to Zhao's going into the Square only on condition that he go too. Looking tired and defeated, Zhao took a bullhorn from one of the students and told them that he "had come too late." He asked the students not to offer themselves in sacrifice, and assured them that "when you end your fast, the government will never close the door to dialogue, never." The students did not realize that this was Zhao's swan song: he was warning them that he could do nothing further to protect them. He had argued in a Politburo meeting at Deng's home on Wednesday that martial law could be only a short-term solution to the questions the students had raised, that it would create more problems than it solved. But he had already lost Deng's support, possibly as much as a week earlier, and could no longer block Li Peng's desire to crush the Movement. Decisions were henceforth out of his hands. And the delay imposed by Gorbachev's visit was over.

Li Peng had a green light. New China News Agency on Friday released a report quoting Li in an interview with Australian Foreign Secretary Richard Woolcott. "Turmoil has occurred in Beijing," Li informed Woolcott. "The Chinese government will, with a responsible attitude, take measures to stop the chaos so as to restore social order." The terminology was not carelessly chosen. "Turmoil" means a situation that is out of control, and "taking measures" means correcting it with force if necessary. It seems that the message had already got to the students. Only minutes before that news release, student leaders decided to end the fast and replace it with a sit-in. They recognized that they needed to regain their strength to face what was clearly imminent now, a military move. Hong Kong Commercial Radio carried a report at midday that two columns of troops, not the 38th Group Army, were making their way toward Beijing. In fact, many more than that were on the move.

It later struck many observers as odd that the military would be brought into Beijing on May 19, just when the Movement was losing momentum. "Things had quieted down by Friday," recalled a foreign scholar. "Through the week there had been an air of electric excitement in the city. People drove around with their banners in the trucks lent by the factories, big demonstrations went up and down the roads, everybody cheered wherever they appeared. But that started to die off by Friday. Some of the buses the universities and the Red Cross had put in the Square for the students had been taken away. I expected they would leave." Still, even she expected some definite move by the government: "something had to happen."

What happened was martial law. The Party's authority had crumbled, and the Beijing police seemed to have let their sympathies slide to the students and could not be relied on to take tough measures against the students. The order was not announced until 9:30 A.M. Saturday, but it was common knowledge in the Square by late Friday afternoon. (The government says that an aide to Zhao Ziyang leaked it to the student leaders.) The move was ill-timed. Student militancy was ebbing. Martial law reversed the tide. A vice-president of one of the universities had gone down to the Square on Wednesday and stayed there for two days and nights, pleading with his students to give up the hunger strike and return to campus. Dialogue, he said, could take place only if the polarization ended. His efforts were wiped out when the news of the government's decision for a military solution was leaked. "He was absolutely furious and upset on Friday by the declaration of martial law," recalled a foreign colleague. "He had just succeeded in persuading his students to go back. Now they refused to go. It was the government's behavior and response to what was going on, as much as the students', that aroused further antagonism and stirred up the movement."

A Chinese participant recalled the same sudden shift. "When we heard about martial law on May 19, we all felt we had to go to the Square. No self-respecting person could stay away. After the martial law decree was proclaimed, the number of people in the Square grew enormously. Everybody came into the Square, from every direction. This was an emergency. Everybody went." The same resolve swept the workers. "Li Peng will never get away with it!" an enraged older worker told a foreign student. "If he brings in the Army, we will oppose him. The people of Beijing will never allow the military in."

According to Article 89 of the Constitution, full martial law can be installed only by authority of the National People's Congress, China's rubber-stamping legislative body. The State Council, which directs the day-to-day work of the NPC under the premier, can, however, impose partial martial law without consulting the NPC. By placing only eight of Beijing's administrative districts (covering Fengtai to the southwest, Shijingshan to the west, and Beijing city proper) under martial law, Premier Li Peng could get what he wanted without consulting the NPC. Had the government chosen to impose martial law over the entire municipality, it would have had to go through the NPC, which, given popular sentiment at the time, would have been a politically awkward task.

Li was arguably within the constitutionally specified limits of his post to sign the martial law order at 9:30 Saturday morning, though the students were quick to challenge the legality of his action. A Beida leaflet entitled "Unconstitutional Regulations Have No Legal Force" pointed out that Article 4 of the regulations governing the State Council requires the State Council to consult with at least the Standing Committee of the NPC in important decisions, and that this was not done. (In addition, there was the delicate problem that the NPC had not yet formally ratified Li's appointment to the premiership.) More broadly, the leaflet insisted that conditions within the capital did not require martial law, and that martial law was overriding other rights guaranteed by the Constitution. The feeling was that Li Peng and a small group within the government had overstepped their powers and were acting unilaterally against the interests of the people.

Late Friday night, Li Peng and Yang Shangkun, representing the State Council and Party Central Committee, respectively, convened a special meeting of leading cadres to announce the government's intention to declare martial law. This was *after* troops had already started to move into the martial law areas. Zhao Ziyang, despite his position as head of the Party and a vice-chairman of the Central Military Commission, was not to be seen. He was invited but declined for reasons of health. The rumor at

People's University that evening was that Zhao had resigned from the government.

Also missing from the meeting were many high-ranking military officers who were in charge of seeing martial law carried out. Contrary to popular suspicions, their absence was not a sign of PLA unwillingness to enforce martial law. It was evidence that popular opposition was being mounted in the streets even before martial law was declared. Crowds near No. 4 Hospital in southeastern Beijing stopped four minibuses carrying these officers to the meeting and held them there until long after the meeting was over.

Li Peng delivered a long-winded speech to the assembled subordinates. Noting the growing anarchy within the city, he said that the Party had decided to take "resolute and powerful measures." He asserted the constitutionality of the decision when he declared that "we will certainly have the backing of the People's Liberation Army, which is entrusted by the Constitution with the glorious task of safeguarding the country and the peaceful work of the people." It was left to President Yang Shangkun to turn to specifics. Yang explained that "to restore normal order there is no choice but to move a group of the PLA to the vicinity of Beijing." He did not come to consult with the cadres assembled before him. They were there simply to be informed that the Army had already entered Beijing "a short while ago." The intention was to get the troops inside the city before the formal declaration of martial law, before the citizens realized that the Army was in their midst and could resist its entry.

Yang insisted that the Army had been brought in "out of absolute necessity. It is because the police force in Beijing Municipality has been unable to maintain order in the municipality. In addition, nearly all the PAP and PSB cadres and police in Beijing Municipality have been working hard day and night for the past month." Yang argued to his captive audience that the Army was being added to the regular security forces not to suppress the students but simply to reestablish basic order in the city. "The arrival of PLA troops in the vicinity of Beijing is definitely not aimed at dealing with students. They have not come here to deal with the students. Their aim is to restore the normal order of production, of life, of work in Beijing Municipality."

President Yang Shangkun was well placed to present the military invasion of Beijing to local cadres. A full PLA general, Yang had at one time been commander of the Beijing Military Region, though his qualifications were administrative rather than military. In May 1989 he concurrently held the posts of first vice-chairman and secretary-general of the Central Military Commission. Yang's ties to the military were personal as well as professional. His younger half-brother, Yang Baibing, was director of the General Political Department of the PLA. Yang Baibing first came to notice in the PLA hierarchy in 1982 as deputy political commissar of the Beijing Military Region, very much through his elder brother's influence.

His directorship of the General Political Department placed him in charge of the Army's political commissars, whose role is to indoctrinate the soldiers. The military modernization program that Zhao Ziyang had been overseeing since 1988 had already reduced the size and status of the commissar contingent within the PLA, and was expected to reduce it further in 1990. Yang Baibing accordingly would have been happy to see Zhao's role within the military reduced. Indeed, martial law gave him the political opening he was looking for. Speaking at a closed meeting at the end of the year, Yang would claim that "if a group of political commissars had not insisted on their political stand and stuck to their positions in times of difficulty, the outcome would have been unthinkable." The Yang family military connections went even further. Yang Baibing's son, Yang Jianhua, was the commander of the 27th Group Army, a major participant in the assault on Beijing. In the days that followed, many would refer with contempt to the 27th as the Yang Family Army.

Still, the man behind it all, the only man who by virtue of being the chairman of the CMC had the constitutional power to order the gathering of military forces around Beijing, was Deng Xiaoping. (The CMC informed the PLA on May 23 that Yang Shangkun had also been given the right to transfer troops.) Deng stayed completely out of public view during these days. After his meeting with Gorbachev on Tuesday, he disappeared and would remain in the background for the next three and a half weeks. Many believed that he went to the central city of Wuhan on the Saturday martial law was declared to bring the military on side, flew back to Beijing, then went down to Wuhan a second time on Monday. One rumor embellished the story by reporting comments he was alleged to have made in a telephone call from Wuhan to Beijing.

Deng's alleged trips to Wuhan probably never happened. Such a visit might have been plausible a decade earlier, when that city was the main military headquarters in central China, but the city's status was downgraded in the military reorganization of the mid-1980s. Furthermore, the men Deng appointed as commanders of the seven military regions in that reorganization were all longtime loyal supporters whose ties to Deng went back to his civil war days as political commissar of the 129th Division and the Second Field Army. Nothing required him to leave the crisis in the capital short of substantial internal opposition within the military, which at this point appears not to have been forming. In all probability, Deng went nowhere. He stayed in Beijing to monitor the progress of martial law.

The night speeches of Li Peng and Yang Shangkun were broadcast over the official loudspeakers to the silent students in Tiananmen Square at midnight Friday. By then, intimidation by words was too late. Li Peng, Yang Shangkun, and ultimately Deng Xiaoping thought martial law was the trump card that would seal the fate of the Democracy Movement.

They assumed that soldiers in the tens of thousands would compensate for the erosion of their authority. But they were wrong. Fate had been changed. Bringing the Army into play now would only sharpen the conflict, not defuse it. And with this miscalculation, the leadership of the Communist Party brought Beijing that much closer to violence.

3

No Place Left
Unguarded
(May 19–23)

At dusk on Friday, May 19, units of the People's Liberation Army attempted to enter Beijing and take control of Tiananmen Square. Martial law would not be declared until the following morning, but the government wanted to steal a march on the democracy activists and have its soldiers in place before the announcement. The people moved more quickly than the Army anticipated. Before martial law was announced, even before the students knew what was happening, the citizens leapt into action. Every military column the Army dispatched—and they came from all points of the compass—was blocked, first by hundreds, then thousands, then tens of thousands of ordinary people.

There must have been a first moment when the first citizens walked out in front of the first truck of the first column of sixty troop trucks coming in from the west along Changan Boulevard just after 8:00 P.M. There must have been a first moment when somone took the chance and decided to risk testing the Army's orders and the soldiers' tolerance. There must have been a first driver who had to decide whether to push on at the risk of running someone down, or to stop and negotiate the situation. There must have been a first commander who saw his column stumble to a halt and could devise no other strategy without the order to go ahead at any cost, an order that would not come. But these moments were followed so speedily by so many other moments of sudden confrontation and stalemate all over the city that it is hard to believe there was a first moment. People arose so quickly in every critical location and in such great numbers that, within a few hours, it was easy to forget that the people's victory started with acts that were spontaneous, unplanned, and without direction. Somehow— perhaps it was the emotional pitch of the hunger strike or the million-strong marches past the Square during the previous week, perhaps it was the students' pleas for citizens' rights—it just happened.

The column coming in from the west was the first of many detachments that would try to move on Beijing from that side of the city. It was stopped on Changan Boulevard just west of Gongzhufen, the traffic circle where the 3rd Ring Road intersects Changan. A forward contingent did make it as far as the circle, but it could get no further. The rest of the contingent was blocked farther west. A team of troubleshooters received orders at 8:40 P.M. to go in and clear the blockade at Gongzhufen, but by the time it arrived, stopped Army vehicles were backed up along Changan for three full kilometers. Forward motion was impossible. Aside from a minor scuffle involving forty-five demonstrators next morning, there was no fighting here.

The regimental column stopped west of Gongzhufen was part of a full division (ten to fifteen thousand soldiers) that had come in from a base in Changxindian, twenty kilometers southwest of the city. They had crossed into the outer suburbs of Beijing over Lugouqiao, popularly known as Marco Polo Bridge. This was the site of the first skirmish in 1937 that led to Japan's invasion of China. On that occasion, as Beijing residents liked to point out in the weeks to come, the invading army declined to subject Beijing to military occupation.

A second column also crossed Marco Polo Bridge. It did not turn north, however, but continued on a northeasterly direction straight through the town of Fengtai to the southwest corner of the 3rd Ring Road. Five troop trucks got past the 3rd Ring Road only to be surrounded near White Cloud Monastery. The rest of the column was stopped back at the Liuliqiao intersection on the ring road. A detachment of PSB police armed with truncheons waded into the crowd at 10:00 the next morning to break up the blockade and succeeded in doing nothing but bloodying one man's head. They were reinforced at noon by a dozen truckloads of armed police. On their foray into the blockade, the PAP injured some two dozen students who did not resist, but the weight of the crowd went against them and they too had to retreat. This skirmish would be the only violent incident between citizens and security forces during this first weekend of martial law.

Twenty kilometers back of this unit at Liuliqiao lay another, much larger column of troop trucks. These soldiers had come from a base near Taiyuan, capital of the neighboring province of Shanxi. Most probably they belonged to the 63rd Group Army. The entire division remained pinned there for three and a half days. Yet another large column of troop trucks came in from Shijingshan, twenty-five kilometers due west of Beijing. These trucks were blocked well outside the city at the town of Gucheng and the Bajiao overpass, twelve kilometers outside the 3rd Ring Road. French journalist Pierre Hurel reported seeing over twenty armored vehicles from the 65th Group Army there on Sunday morning, immobilized near the large Capital Iron and Steel Works, the oldest large industrial enterprise in the Beijing region.

The units that approached Beijing along routes from the west and southwest constituted the main bulk of Friday night's invading force. But the Army also tried to make its way into Beijing from the south, east, and north as well. A contingent of troops came up from Nanyuan, the site of a civilian-military airfield south of the city. Again, six troop trucks in the forward part of the contingent were able to push ahead and made it as far as Yongdingmen (Gate of Eternal Stability), the intersection at the southern boundary of the old city. A foreign observer watched them withdraw Saturday morning, "the young soldiers standing in the back grinning sheepishly with relief, waving and flashing V signs to the passing crowds." The main body of this column was stopped by students further south at the outlying town of Donggaodi. They were observed the following morning by the same foreigner. "I counted forty-seven armored personnel carriers and thirteen trucks filled with troops. Despite the fact that they had been stranded in their vehicles for over twenty hours, the troops seemed well disciplined and impervious to the sporadic student harangues." Lest this column make a second attempt on the Square directly north, a sturdy barricade of two buses and five trucks loaded with bricks was placed across Qianmen Street, the street leading north from Yongdingmen to Qianmen (Front Gate), the southern approach to Tiananmen Square.

Convoys of troop trucks and armored vehicles also approached from the east a few hours later, some from the Sanjianfang Airfield, the rest from military camps in Tongxian further east. All were stopped at the three main intersections on the east section of the 3rd Ring Road, Dabeiyao (where it crosses Changan Boulevard), Hujialou (to the north), and Shuangjing (to the south). Popular resistance was in full flood at Dabeiyao by 3:00 A.M., when a foreigner sleeping nearby was woken by the noise of people rushing to protect Beijing's eastern flank from the Army. An articulated trolley bus stretched across the road stopped the truck convoy, and then workers dumped three cartloads of gravel behind the wheels of the last truck after it stopped moving. The trucks were pinned down, unable even to retreat.

Journalist Pierre Hurel came upon the convoy an hour later. A worker standing on the hood of a truck played "The Internationale" on a trumpet while a student on the roof of a bus in the barricade exclaimed: "Who is causing this unhappiness for the people? Who wishes to turn us from our national destiny?" According to Hurel, the convoy consisted of "four trucks armed with water cannons, six trucks carrying enormous storage bins the size of a missile that contained tear gas canisters, and eight troop transports."

What Hurel and the crowd took to be tear-gas transport trucks were not. If the storage bins made Hurel think of missiles, that is because missiles are what they contained. A foreign military observer also at Dabeiyao was able to identify these trucks as antiaircraft resupply vehicles. He was amazed by his discovery, but kept the information to himself so as not to fuel the crowd's anger. The presence of an antiaircraft unit on the edge of the city is disturbing. What could have been the logic of including this sort of weaponry in a martial-law force—except in anticipation of conflict between military units, a possibility no one as yet even dreamed of?

Hurel continued another ten kilometers east of the city and found a second group of about twenty troop trucks blocked by eleven trucks from a nearby cement factory. This contingent belonged to the same unit as the vehicles farther in at Dabeiyao. Again, crowds had broken up the column at several points and none of the trucks had gotten through. On one of them someone had written: "The people's soldiers are not guard dogs for the emperor's house," a veiled reference to Deng Xiaoping. "Even if they wanted to retreat," a fifty-year-old worker told Hurel, "we are going to keep them here until noon. The people of Beijing must see what the soldiers have come to do in the capital."

At Hujialou, the next intersection on the 3rd Ring Road north of Dabeiyao, another convoy coming in from the east was also stopped about 3:00 A.M. According to New China News Agency, which reported the incident, an old lady lay down on the road in front of one of the trucks. "If you want to move," she declared, "drive over me."

The *People's Daily* printed a picture of the truck column at Hujialou in its May 22 edition. The report quoted an officer as stating that "since we are the people's own army, we will never suppress the people." It is hard to tell from the comment whether he was making a commitment not to oppose the wishes of those who had stopped him, or indulging in the formal rhetorical trick of distinguishing "the people" from "hooligans," who by definition had to be suppressed. But this confusion is always ambiguously present in the language of governments or armies that declare they represent "the people."

Several Army units also tried to penetrate Beijing from the north, though reliable information on military movements from that direction is scarce. We do know that a large column set out from Changping, a road and rail nexus thirty kilometers beyond the northern suburbs. Near the Summer Palace, the baroque pleasure park built by the Empress Dowager at the end of the last dynasty, people lay down across the road and halted part of this column, roughly three to four thousand troops. Another column of troop trucks was stopped west of the Summer Palace at Wenquan. The PLA was also blocked at Qinglongqiao, the next rail station north of Changping. A Beida student at the Qinglongqiao blockade tried to find out where the soldiers had come from, and what they knew, by talking to them. "They had come in from Zhangjiakou, a large railroad and military base northwest of Beijing. The soldiers didn't say much—they weren't allowed to. So we couldn't find out exactly where they had come from. They must have originated from at least as far away as Datong." Datong is a district military center in neighboring Shanxi province two hundred kilometers southwest of Zhangjiakou on the same rail line. "We tried to speak to the officer in charge, but he didn't seem to know what was happening. Even if he had known, he had to stand on the government's side and couldn't reveal his attitude. He was only carrying out orders."

In addition to these units, an armored column also approached Beijing from the north but retreated the following morning to a position farther outside the city. One report says this contingent included sixty-five APCs and seventeen tanks. The sighting of tanks cannot be confirmed; tanks appear not to have been part of the force mobilized for this operation.

By the time the sun rose on the morning of Saturday, May 20, at least seven divisions (and probably more) had attempted to enter the city under cover of night by way of at least six main routes and several minor ones. Roughly five of these divisions included armored vehicles, most if not all APCs, used not to clear a line of advance but simply as troop transports. These forces were followed Saturday morning by a smaller second wave of motorized units that did no better. Despite an urgent order issued at 6:00 A.M. by the Beijing Municipal Party Committee calling for mass mobilization to assist the Army, this wave had even less success getting into the city. Within barely twenty minutes of setting off from its base in the Fragrant Hills beyond the Summer Palace, for instance, a truck column in the second wave found its way through the western suburbs blocked by tens of thousands.

In almost every case, the incoming columns were stopped midroute not by students but by ordinary citizens who chose to defend their city against a military occupation. As a Capital steelworker explained to a *New York Times* reporter on Saturday morning, "We have towels for tear gas and maybe buckets of cement to make road blocks, but besides that, we come just as we are—people." Against simple resources of numbers and non-violence, and without the order to go ahead at any cost, the Army was powerless to proceed.

The students in the Square did not know that the citizens' defense would succeed so well. They expected the Army to get to them; they expected to be removed from the Square by force. And so they sat there patiently waiting Friday night for an army that never came.

Nor did the Army anticipate failure. The Martial Law Command clearly intended that its forces reach the Square. Robin Munro, senior research associate for the New York-based human rights organization Asia Watch, spent the early hours of Saturday morning on the Square with a plainclothes officer from a military unit based in Shaanxi, probably the 47th Group Army. The officer was one of numerous scouts the Army kept posted in the Square through the entire operation. He knew the precise routes the various columns were taking into the city, and was on the Square waiting to coordinate their arrival. By 4:30 A.M., after student loudspeakers had made a series of announcements that military columns had been stopped at every point they tried to enter, the officer had to admit that the targeted time for all the convoys to converge on the Square had passed. The unexpected had happened.

(Undercover agents remained on the Square throughout the weekend. One even took over the microphone of the students' loudspeaker system Saturday morning and announced: "I have good news for you all. Our People's Liberation Army comrades have come to help keep order. Let's welcome them." No one supported the suggestion. A student leaflet that was passed out to troops on Saturday did say that up until the Friday night offensive, the students had hoped that the PLA would come out in support of their demands. But by midnight Friday the students knew better than to think that the PLA had shown up at the gates of the city to champion the call for democracy against a recalcitrant Party.)

The Martial Law Command did not know what to do in the face of this determined and universal opposition. It attempted to get a sense of the overall situation early Saturday morning by sending helicopters over the city on aerial reconnaissance. An eyewitness at the Jianguomenwai foreign diplomatic compound saw "five helicopters flying in formation. They were flying the same height as the eleventh floor of the diplomatic apartment buildings. They had civilians on board, plus the military pilots, and they were obviously doing a reconnaissance of the area to figure out what was going on." The civilians were presumably government or Party officials. The helicopters made another reconnaissance flight across Tiananmen about noon. Helicopters also flew over the western part of the city later Saturday afternoon.

The noon flyover was a propaganda as well as reconnaissance mission, for the helicopters dropped leaflets on the Square. One rumor that got going around the city was that these were Air Force rather than Army helicopters, and that the Air Force was on the people's side. A crowd in the eastern part of the city actually cheered the helicopters as they flew over, interpreting their presence as a gesture of solidarity with the demonstrators. Rumor even suggested that they were dropping money on the Square in support of the Democracy Movement.

For those who were in the Square, the mood was less sanguine. At 10:00 A.M., the government loudspeakers at Tiananmen started broadcasting the martial law orders and warned that "PAP and PLA soldiers have the right to exercise any force necessary to stop or prevent any violation of martial law orders." The same message was printed on the leaflets being dumped from the helicopters. "The leaflets warned people that they should get off the Square. We feared that there would be a military repression that night," recalled a foreign scholar who was there on Saturday. It was the act of dropping leaflets itself, rather than the message, that disturbed some. "A long-term Party member visited a foreign friend that afternoon. "I still remember his comment: As far as he knew, the technique of dropping leaflets had only been used twice before in modern Chinese history to communicate with the people. Once was by the Japanese when they were invading, and once was by Chiang Kai-shek under the Nationalists. His own position was that he couldn't go out and march himself, but he encouraged all the young people in his section to go out." The technique was conduct unbecoming to the party he supported.

The helicopters gave many the impression that the Army was on its way. A foreign-language teacher who was in the Square that afternoon reported just this. "The students there were certain that the Army was coming in. The helicopters had been going over, and the Beishida students we spoke to said that they expected the Army to come in after dark. They had finished their hunger strike and started eating soup so that they would have the strength to face the soldiers. They intended to stay. But I felt intimidated by the possibility and left just as it got dark."

Whatever anxiety the students felt regarding an impending attack was more than compensated for by the heady feeling that the Army had been defeated by people power. The people believed they controlled the city— as indeed they did, for the moment. "It was fantastic to see the manner in which the populace of Beijing went into action for the students," remarks Canadian scholar Bruno Munier in his diary. "It seemed as though the entire population of the city had gone down into the streets." By noon on Saturday, the citizens and students had organized a counterauthority within Beijing against a military invasion by their own government. "Beijing was no longer obeying its government; and although proclaimed, martial law was given absolutely no respect."

The students capitalized on the citizens' victory and quickly took over the work of directing the city's defense. The citizens accepted their leadership without question, for these young literati were the voice of the nation's conscience. Coordinating the work of holding the Army out of the city proved to be an enormous task. Much had to be left to chance and circumstance, but the popular support the students' cause enjoyed meant that chance and circumstance succeeded quite well. The students were helped to some extent by information that friends and family in the upper levels of the Beijing bureaucracy were feeding them, though not all of it was reliable. Student leaders had to negotiate their way across a wide flood of rumors and uncertainties. Most decisions had to be made on the basis of what was going on in the streets.

While some students stayed with the barricades and monitored developments there, others traveled around the city in trucks spreading the latest news (laced with heavy doses of unconfirmed rumor) and distributing leaflets to citizens and soldiers alike. They held councils at central locations to coordinate the work of blockading the Army and to discuss strategy, and relayed their strategies back to the blockading positions by telephone, by walkie-talkie, and by couriers on bicycles or motorcycles, or even in ambulances.

The motorcycle brigades, known as the Flying Tigers, were a showy part of this communications operation. A motley collection of Hondas, domestic motorbikes, and mopeds, the Flying Tigers represented the contribution of Beijing's collective and individual businesses to the Movement. Petty merchants were stigmatized in imperial times as men who pursued profit rather than learning, the only true undertaking of a gentleman. They had largely disappeared under Mao Zedong's rigorous Communism, but were flourishing again thanks to Deng Xiaoping's program of economic liberalization. The old social stigma against traders still remains, fueled by jealousy over how well private entrepreneurs have done in the past decade while everyone else felt the pinch of inflation. Small-time businessmen, though, had their own complaints about the corruption of licensing officials, and for this reason came to the students' support with money and motorcycles.

To keep the street situation favorable to their cause, student pickets worked hard to maintain public order, both in the standoff zones and throughout the city generally. Sometimes working with PSB personnel but more often alone, they strove to ensure that their nonviolent movement not lose that one essential ingredient that pulled the fangs from the Army. PSB on duty within the city gave the appearance of not wishing to obstruct the Democracy Movement.

Globe and Mail reporter Jan Wong talked to a senior PSB officer as he ate a nighttime meal of soup at the edge of the Square during one of the first nights of martial law. "We'll never hurt the students," he told her. "We'll never beat the students. We think they're justified. We're here to prevent"—he paused—"turmoil, or something." When asked what he made of the citizens' successful attempt to hold off the Army, he replied, "Well, I can't say." But his tone was entirely sympathetic.

Out on the edges of the city, the blockaded soldiers were overwhelmed and disoriented by the scale of popular opposition they encountered. They were unprepared for this resistance. A Chinese doctor who went west out Changan Boulevard at 2:00 A.M. Saturday to look at the convoy there remarked that "at that time, the soldiers were quite polite. They showed no aggressive behavior. They went wherever we pushed them." He went back after dawn and joined in the propaganda work that citizens were doing to explain to the soldiers why the people had stopped them from entering the city.

"You shouldn't be here," someone in the crowd said. "This is not right."

"Do you know what has been happening in Beijing?" the doctor asked a lieutenant who was climbing down from his truck.

"We don't know," he replied.

"Then why have you come?"

"We were ordered to take up positions around the city."

To the ears of Beijing people, their accents sounded southern, so they asked some Hong Kong students in the crowd to explain to the soldiers why they shouldn't be doing this.

"Do you know what has been going on in Beijing?" they asked. "Do you know why students have been fasting? Do you know what they're fasting for? You should know. Haven't you read the newspapers?"

"We haven't seen a single scrap of news for a week," they replied.

"Even if ordered to, you must not shoot."

"On no account will we fire on the people," the lieutenant vowed. "Even if we are ordered to shoot."

"You shouldn't go to Tiananmen," someone in the crowd repeated. "Go back."

"We have our orders. We can't go back. But as long as you block us, we will not go forward. We won't run anyone down."

The talk of shooting implied what everyone assumed: that the convoys carried arms, even if they had not yet been distributed into the soldiers' hands. The outward appearance of most of the troops throughout this weekend was of an army without weapons. Only occasionally were soldiers seen with rifles in their hands. The casual manner in which they were held suggested to one observer that they were not loaded. Many units never showed any guns. Although they had them, they kept them carefully stashed away. Even when guns were seen or mentioned, it was never suggested that they would be fired. At this point, the Chinese government was not brandishing the threat of violence at the people of Beijing. Nor did it want the world to get the impression that it was prepared to use bullets to get the students out of Tiananmen Square.

The guns that people spotted were the PLA's standard combat issue, the Type 56 assault rifle. This is China's copy of the Soviet AK-47, and is usually referred to as such. The AK-47 is an automatic rifle designed for assault purposes. It is accurate within a range of four hundred meters, fires at a rate of about ninety rounds a minute, and causes particularly nasty wounds (the bullets do not simply pierce the human body but tumble as they hit a target). It is a weapon of warfare, not of civilian control.

The people were successful in keeping the soldiers out of the city throughout the day on Saturday. But that night, tension mounted with the anticipation that the Army might discontinue its passive response to the obstruction and make a new push toward Tiananmen. While crowds kept the besieged forces under scrutiny, thousands more volunteers stayed awake and posted themselves near other major intersections around the city in case the Army made new moves. The citizens were particularly alive to the possibility that the Army might try to infiltrate soldiers into the city. Infiltration was the only way to proceed after the failure of the open operation the night before and with the city now under vigilant guard. This is precisely what the Army tried to do. On Saturday night, soldiers moved about the city dressed in civilian clothes (though their close-cropped haircuts and the imprint of military capbands on their foreheads often gave them away). They offered false identities when they were challenged. Some used ambulances as transport, though even ambulances were stopped and searched. About midnight at the Xinjiekou intersection on the west side of the city, some four thousand people stopped an ambulance ferrying soldiers down to the Yongdingmen Railway Station. The soldiers were easily identified when citizens discovered that they were carrying their uniforms in paper bags. One of these soldiers was able to escape by asking to go to a public washroom, then climbing out a back window and taking off down a dark alleyway.

Even outside the city, such deceptive tactics had to be used if the Army wanted to move soldiers into more effective positions. A regiment based in Tongxian that had been turned back Saturday morning set out again at 11:30 Saturday night. Wearing civilian clothes and traveling in unmarked buses, their assignment was to set up a forward base in No. 292 Hospital, a PLA facility halfway between Tongxian and Beijing. These soldiers were armed. To get through the barricade at Tongxian's Beiguan Bridge, where several hundred soldiers and policemen were pinned down by a large crowd, the plainclothes soldiers flashed victory signs to the crowd and hurried through in the wake of a vegetable delivery truck that had gotten permission to pass the blockade. The crowd had no idea these were soldiers. By avoiding the main roads and traveling on small country sideroads, the contingent was able to get to No. 292 Hospital unopposed.

Small victories of this sort, however, did not substantially shift the balance from the citizens to the Army. What people feared was a more spectacular attempt to send in soldiers in large numbers. At one point Saturday night a rumor went through the Square alleging that the PLA would try an air assault. The students fanned out to prepare for parachutists, and some took to flying kites high over the Square to discourage helicopters from hovering in overhead. No such attempt was ever made. It was a false alarm. The same rumor was repeated Sunday afternoon, though the landing site would not be the Square but the Temple of Heaven in the southeast quarter of the city. Beida and Renda students mobilized workers to block the gates to the Temple grounds with trucks, but the anticipated jump never materialized.

The main PLA offensive Saturday night may have come neither by land nor by air, but underground. A giant network of tunnels connects the downtown core with the city suburbs. This labyrinth was the result of Maoist guerrilla warfare transplanted to modern urban life. Following the major rupture of Sino-Soviet relations in 1960, when China chose to turn its back on its imperious senior partner, the Chinese government invoked the specter of Soviet aerial bombing to revive its sagging popularity after several years of economic mismanagement. The entire population of Beijing was dragooned into volunteering labor to build this network of escape routes underneath their homes and offices. Beijing's subway, which opened in the mid-1970s, is linked into this system. The main tunnel of the underground defense system is wide enough for two trucks to drive abreast. It connects the Great Hall of the People, the center of state administration on the west side of Tiananmen Square, with the Beijing Garrison Command in the Fragrant Hills. Side tunnels branch to other government buildings. Further out are located entrances in the countryside at which elderly peasants are posted as watchmen. The existence of the tunnels is classified information, but it is also common knowledge. What might serve in an armed struggle against a foreign invader thus will not work against the people of Beijing. It is simply too easy for the people to find out if something major is happening under their feet.

In previous weeks, troops had already been moved in through these tunnels in small numbers to the major government installations around the Square: the Great Hall, the History Museum, the Forbidden City, and Zhongnanhai. The plan for a night attack during the first weekend of martial law—and the only contingency plan the Army seemed to have once its above-ground attack was frustrated—was to move a large contingent of troops by truck through the main tunnel and take the Square from the government buildings surrounding it. The subway was shut down at 10:00 A.M. on Saturday, the moment martial law went into effect. People assumed that this was so timed to allow soldiers to be moved underground, and they responded by immediately blocking the exits. At the station down at the south end of the Square, supporters drove public buses right into the subway exits to prevent soldiers from pouring out en masse.

By report, the Army moved several divisions through the tunnels that night. One of the watchmen out in Haidian District, where most of the universities are located, detected the movement in the main tunnel. He immediately contacted the students, and several thousand poured down into the tunnel to stop a similar number of troops. The soldiers could do nothing but turn around and withdraw. Infiltration by tunnel failed.

The only soldiers to make it into the city that night came by rail. A ten-carriage train of over a thousand troops armed with automatic weapons pulled into the Beijing Railway Station in the early hours of Sunday morning, only to find themselves surrounded by tens of thousands of vigilant citizens. A thirty-car train arrived in its wake at 6:13 A.M. It had managed to escape detection all the way in from Shacheng Station, almost a hundred kilometers northwest, by appearing to be an ordinary freight train, although in fact the open freight cars were transporting the soldiers of an artillery unit. According to foreign journalists' reports, the soldiers trapped in the train station came from either Shaanxi or Sichuan province; according to a later government publication, one of these units came from rural south China.

The PLA made other attempts to move soldiers by train during Saturday and Sunday nights, but to no avail. Train stations in fact proved to be weak links in the PLA's troop transport system because a handful of bold protesters could tie up thousands of soldiers by sitting on the rail lines leading in and out of stations. Troop trains were blocked Saturday night at the suburban stations like Shahe, fifteen kilometers north of the city, and Changping, another twenty kilometers out. They were also blocked at far greater distances from Beijing, though the only information I have comes from the next night. One troop train routed to Beijing from the northeast was stopped at 1:00 A.M. at Qianan Station, two hundred kilometers due east of Beijing. The unit's commander was told to "await further orders" but heard nothing for two days. Troop trains were not uniquely vulnerable, however. Even regular passenger service was disrupted. At the Jinzhou Station, another two hundred kilometers east of Qianan, a group of Jinzhou University students lay down on the tracks and refused to move. The train they were blocking, the #290 via Luanxian to Beijing, carried no one in uniform, but the students suspected—rightly, as it turned out—that some of the passengers were soldiers. The PLA was trying to infiltrate soldiers into Beijing by having them dress in civilian clothes and travel incognito on regularly scheduled passenger trains like the #290.

As dawn broke on Sunday, May 21, the citizens of Beijing found that they had passed another night successfully defending their city. Confidently, even joyfully, they continued in this task. "Within Beijing, the populace exercises very strict control over entry into the city and at the principal intersections," reported Bruno Munier. "If order reigns, it is in the absence of a single policeman." Munier drove north out of Beijing that day to see the Great Wall seventy kilometers north of the city. "The last control of this type that we encountered was situated about thirty kilometres outside the city. From all the evidence, Gandhi would have been happy to see what is unfolding here, because everything is going calmly." On his return trip he encountered neither military checkpoints nor troops. The only controls on access roads into Beijing were manned by the people. "It is incontestable that order reigns, but it is not the order of the government. In fact, more and more I have the impression of being, in Beijing, no longer in a people's republic, but in a true People's and Students' Republic. For the moment, the effective government is principally the Autonomous Students' Federation."

Street barricades restricted movement within the city on Sunday, but no one objected to the inconvenience. A foreign teacher traveling across the city on Sunday found barriers at every turn manned by thousands of people. "The concrete road dividers made Ss along the road. There were buses at all the intersections, sometimes with flattened tires. There were no identifiable leaders, but everybody knew what to do. They were all working together in concert, yet I couldn't see anybody taking charge." The largest barricade was in Fengtai in the southwest suburb. "It consisted of two giant concrete sewer pipes across the road, plus four rows of trucks in the middle of the road, and rows of trucks three or four deep in each bicycle lane."

Military convoys on the city's perimeter continued to be immobilized through Sunday by the masses of citizens determined to keep them out. In at least one instance on Sunday morning, citizens took the initiative in their struggle with the military. A crowd of over two thousand people went down to the military airfield at Nanyuan, seven kilometers south of the 3rd Ring Road, and blocked the entrance. Nanyuan was the main airport on which the PLA relied for airlifting supplies into the Beijing region. The crowd prevented oil tankers from getting into the airport, thereby cutting off the supply of air fuel.

The only movement of soldiers during Sunday was the assignment of small contingents of soldiers to "guard" key public buildings, ministries (notably the Ministry of Broadcasting and Television), and media offices (including the *People's Daily*). The *People's Daily* compound in the eastern part of the city was slated to be occupied midday Saturday, but the regiment assigned this task never made it out of its camp in Tongxian. The following day, however, the staff was surprised to see a convoy of Red Cross ambulances carrying white-robed medical personnel drive into the newspaper compound. That was the only way the PLA contingent assigned to *People's Daily* could manage to get inside the city. Other units tried other disguises, but even these precautions did not always guarantee that they would go undetected. One group of plainclothes soldiers coming in from Shunyi county on Sunday morning with orders to occupy the Color Television Center got caught out when a suspicious citizen noticed that their undershirts were Army issue. Once inside the media compounds, however, soldiers did not conspicuously interfere with operations. Staff in some places were uncooperative, but they did not openly resist the military's presence. Resistance at newspapers took the more subtle form of declining to publish pieces by PLA authors.

On Sunday, the propaganda war between the people and the government was in full flood. Because neither side was completely successful in getting the official newspapers to present its views (reporters were resisting the official view whenever they could), both turned to the time-honored tactic of printing and distributing leaflets. Government leaflets came by air, dropped by helicopters over Tiananmen Square and some of the university campuses.

"A Notice to the People of Beijing from the Martial Law Command of the PLA" was dropped on Sunday. Its opening sentence attempted to rewrite history by saying that soldiers had been sent into Beijing at 10:00

Saturday morning in order to impose martial law. That was simply the hour that martial law was declared; citizens had been pinning down the soldiers in the suburbs at least thirteen hours before that. The leaflet went on to list four "announcements" from the Martial Law Command. It said the Army had been sent in "solely for the purpose of protecting the security of the capital and restoring normal order, and not to deal with the patriotic students." In the wake of the standoff, the Martial Law Command warned that "the martial law troops must resolutely carry out the orders of the government. We have the duty to take any effective measures to reverse the present situation." It insisted that the Army supported the demands for an end to official corruption, for "socialist democracy," and for a proper legal system. Martial law was simply intended "to create the sort of social environment favorable to implementing these demands." Finally, it promised that although the soldiers would observe discipline, they would also "oppose any acts or words that could damage the relationship between the army and the people that is as close as fish to water." The leaflet closed with a politely worded request for an attitude of restraint on the part of city residents and students.

Another Army leaflet dropped on the Square and Beida on Sunday and Monday tried a simpler approach, listing a series of twenty numbered slogans that supporters of martial law should use. Here the Martial Law Command was imitating the student organizers, who also printed up sheets of numbered slogans and distributed them to demonstrators so that they would be speaking with one voice. The Army vainly hoped that the people of Beijing would rise up in vocal support of martial law. Its list included such slogans as these:

No. 4. "We Don't Want a Repeat Performance of the Tragedy of the Ten Years of Chaos!" (The government found it useful to play on fears that the Movement would lead to another Cultural Revolution. The visual parallels of students marching under waving banners were obvious; but the calls—"permanent revolution" versus "democracy"—were polar opposites.)

No. 5. "We Resolutely Support the Important Speech of Li Peng!" (This was Li's eleventh-hour speech before city cadres on Friday night. Protesters in fact were parading around the city that night in trucks emblazoned with banners that said "Down with Li Peng!")

No. 13. "Workers, Stay at Your Jobs and Devote Yourselves to Production in Keeping with the Spirit of Being the Masters of the Country!" (The workers had a different notion about how to express their mastery of China.)

No. 14. "Students, Return to Classes Immediately and Maintain Normal Educational Order!"

None of these slogans passed the lips of Beijing people that day. Instead they chanted the slogans that students printed up and distributed, slogans voicing quite different aspirations:

No. 1. "We Want Democracy and Rule of Law, Not Guns and Truncheons."

No. 19. "The Party is not the Li Family Party, the Army is not the Yang Family Army," referring to Li Peng and Yang Shangkun.

No. 21. "Military Control Brings Chaos, Martial Law Induces Paralysis." (This was in response to the government's charge that the students' demonstrations had paralyzed the city.)

No. 22. "When Officials Oppress the People, the People Cannot Survive; When the Army Oppresses the People, the State Will Topple."

No. 28. "We do not Accept that the Gun Rules the Party and the Government!"

In addition to lists of slogans, student presses that weekend printed up a plethora of appeals addressed to the soldiers. One of these, printed at Beida and handed out on Sunday to soldiers in the southwest quarter of the city, warned the soldiers that they were being hoodwinked by "a small group of bureaucrats represented by Li Peng who, fearing the strength of the people and opposing reform, have placed the Party secretary-general [Zhao Ziyang] under house arrest and moved large numbers of troops in from other regions in the hope of suppressing the people and the students." The leaflet assured the soldiers that "we the citizenry still have deep feelings for the people's army. We do not oppose you at all, nor do we hate you. What we oppose are these reactionary bureaucrats who hope to fool you." The text closes with an expression of solidarity with "you patriotic officers and men who have come from afar only to discharge your military duty."

Another Beida leaflet chose to remind the soldiers of the history of student movements: "The students are guilty of no sin and have done patriotic service. Seventy years ago the northern warlord government suppressed a student movement [May 4th], thereby hastening its own demise. Thirteen years ago the April 5th Movement in front of Tiananmen [in 1976] was suppressed, but even so the verdict on it was very quickly and thoroughly reversed." The students wanted the soldiers to remember that the long-term victories belonged neither to the Army nor to the government it served, but to the students. They also wanted the soldiers to think about their own oppression. "You come back from the dangers of frontline service and just look at these corrupt old farts! Their pores drip with oil and they squander the blood and sweat of the people. They seize the fruits of victory paid for with the blood sacrifice of fighters on the frontline. Seeing all this, are you not righteously indignant? So now who is looking after whom? Who is looking after the hunger and suffering of ordinary people? With whom should accounts be settled?" The government's decision to bring in troops, the student authors of the leaflet insisted, was a "disgrace to the Chinese Communist Party and a "great scandal in contemporary international society." The students appealed to their "brother soldiers" to act as "defenders of the heroes of democracy."

Student organizers also handed out leaflets to demonstrators to help them work effectively in their standoff with the Army. One of the most interesting is an open letter from a self-styled "old soldier" to the students, printed up and distributed the next day. It offers practical advice for countering a military offensive on the city:

> Several hundred thousand people in Tiananmen Square is not a problem that a small number of troops can resolve. Martial law has been imposed for dozens of hours and yet no action has been taken against the Square. This is because the military and police presence is insufficient. Soldiers and police-men receive different training; it is very difficult for the latter to point their guns at the people. Ten million people bitterly opposed to a common enemy cannot be countered.

> Tactics:

> 1. Intersections: For the masses to take it upon themselves to block military vehicles is an unprecedented gesture. On no account must you withdraw. Even if you have heard that one or two units have already entered from another direction, you still must not withdraw. This is because one or two units are not enough to solve the problem: they cannot carry out martial law or impose military control on the Square. Staunchly defending the intersections spells victory. To achieve this aim, a stable command structure must be set up at every intersection, improving the present state of disorga-nization.

> 2. Break-up: This is a military tactic, but the people can use military methods to control the military. When completely prepared, you can let through two-thirds of an army and block the last third. At the next intersec-tion, let through another two-thirds and block a third. By breaking up a well organized army into separate parts so that the head cannot see the tail, the soldiers' morale and combat strength will be fundamentally undermined.

This was a technique the people would use to great effect two weeks later.

> 3. Gunfire and tear gas: Firing guns or letting off tear gas against over a hundred thousand people crowded together will inevitably lead to chaos and a large number of casualties. You must warn the military leaders that who-ever gives such an order will face court-martial and be executed. Beijing may be temporarily brought under control by massacre, but how can the dozens of major cities throughout the country be brought under control? We don't have that many soldiers.

> 4. You must believe in the basic quality of the people's soldiers. Do careful ideological work with them. Don't just tell them not to come into the city: tell them to turn their guns and stand on the side of the people.

Ideological work seemed to win the demonstrators quick dividends. Soldiers certainly did not contemplate "turning their guns" against their leaders, but they did give promises of nonbelligerence. Student propa-ganda made inroads among the martial law troops in part because the soldiers had practically no idea of why they had been sent to Beijing.

Officers at the divisional level and above seemed to have some idea of was going on, but those at the regimental level and below, and even more the rank and file, were in the dark about the entire affair. They lacked knowledge of the situation they were being sent into. Nor were they prepared ideologically. The only propaganda that Army commissars seem to have given the soldiers was the hotly disputed April 26th Editorial.

To those who blocked them, the soldiers' ignorance was baffling. One foreign eyewitness asked a soldier stranded in a truck southwest of Beijing what he was doing there.

"I don't know," he replied. He asked him if he realized what had been happening in Beijing.

"Well, we haven't seen a newspaper or television for a week, since before martial law."

"Do you normally get newspapers and television?"

"Normally, yeah, but they just told us to get in these trucks and go to Beijing. They didn't tell us what was supposed to happen when we got here."

The soldiers knew only that martial law had been imposed. They didn't know what that meant, or what it was supposed to achieve, or what they would be expected to do once they got off their trucks. Most of them never found the opportunity to get off their trucks. Without orders to go ahead at any cost, the soldiers were stuck—and vulnerable to the people's pleas that they not suppress the students, and that they not use weapons against civilians. When a student from Qinghua University asked a soldier in the west end of the city whether he would fire on the students, he replied, "I will absolutely not use my weapons. The students are like my brothers and sisters. Even if I'm ordered to shoot, I'll just fire into the sky."

On Sunday afternoon, three student representatives received more guarded assurances from a regimental political commissar named Liu Zhi-jun, whose unit was among those marooned around Beijing, according to a New China News Agency report released three days later.

"The PLA is led by the CCP," said Commissar Liu, "and we will comply with the Party's command. The Party Central Committee and the government have given explicit instructions to the troops not to open fire and to do the utmost to prevent bloodshed." PLA conduct through the weekend confirms that such an order was in place. To underscore his message, Liu assured the student representatives "that the troops will not fire on the students." He added, however, that "we will by no means be soft on those evil elements who are bent on opposing the government with sabotage activities." In other words, though the people might count the soldiers' passivity as their victory, the Army's resolution not to use weapons against the people came not from the popular opposition, however impressive, but from the political leadership. Individual soldiers may have flashed victory signs at the people, but none broke ranks and joined the protest.

The one sign on Sunday that ranks might be breaking at the top of the military structure was an open letter to Deng Xiaoping in his capacity as chairman of the CMC signed by seven retired generals. The letter asked that the PLA not be used against the people. The generals stressed that it would be better not to have the PLA push ahead and enter the city, that such an action could only make a grave situation worse. Their opposition appears to have arisen because of—or at least to have been expressed in terms of—what they considered Deng's irregular procedure in calling out the Army for domestic duty. Several members of the Central Military Commission were not even informed of his decision. This letter was leaked to the students and appeared on wall posters around the city. The signatories made an impressive list: a former minister and vice-minister of defense, a former chief of the General Staff, and four full generals. The letter was said to have been signed also by a hundred lower-ranking military officers, although their names do not appear on any of the copies in circulation.

Seeking to confirm, and reinforce, this opposition to martial law from within the higher echelons of the military, eleven students of the Chinese Science and Technology University went on Sunday evening to the residence of retired Marshal Nie Rongzhen to solicit his views. Marshal Nie had commanded the forces that captured Beijing from the Nationalists in January 1949 and was perhaps the most honored military man in China forty years later. Nie asked the students to vacate the Square and assured them that the Army would not be used to suppress their movement. Another delegation of seven students called on another prominent Army man, Marshal Xu Xiangqian, and though they did not succeed in meeting with him, received a similar response from a member of his staff. State television that evening reported the messages of Nie and Xu, both loyal supporters of Deng and Yang, that the Army would not fire on the people. It mixed this message, however, with another from the vice-mayor of Beijing, who appeared on television in military uniform to warn that barricades must be dismantled and vigilance heightened lest chaos take hold.

Fear of a renewed military offensive rose again Sunday night despite the reassuring comments of the two PLA marshals. Rumor said that the Square would be cleared out at 5:00 A.M. Monday morning. Not only that, but word was that the government was prepared to sacrifice as many as 300,000 lives. So widely was this rumor believed that state television felt compelled to deny it on the Sunday evening news. Many students and citizens were convinced nonetheless that the Army would arrive and went out to bolster the city's defenses. "All the thoroughfares were completely blocked off," recalled a Chinese doctor. "People went of their own accord. It was not organized: people just came out on their own initiative to help the students. They put up all sorts of barricades. The traffic dividers running along Changan Boulevard were pulled out across the road. Buses were taken out of service and placed as barriers at intersections, which were defended by citizens or students. It was like this everywhere in the city."

About midnight, two men came to the Square to warn student leaders that the Army would push through to the Square at dawn, and asked them to arrange for the students to withdraw so that no blood would be spilled. Significantly, one of these men came from the China Handicapped Association. This organization has for many years been headed by Deng Xiaoping's son, Deng Pufang, who was crippled by Red Guards during the Cultural Revolution. Regarded as having made a fortune through official profiteering, Deng Pufang was satirically known as the "crown prince." A poster went up during the Movement warning him that if he didn't change his corrupt ways, he was likely to lose his other leg as well! The messengers may have been using the association's link with Deng Xiaoping as a way of assuring the students that their authority came from the highest levels: an informal but official source.

After much discussion and agonizing, Beishida student leader Wuerkaixi decided to trust their report. He argued within the student leadership that they had scored a victory and that it was now time to vacate the Square lest the Democracy Movement turn into a bloodbath. He was opposed. Other leaders argued that this was nothing more than a stratagem from *The Art of War*, a government trick to fool the students into leaving without a fight. Wuerkaixi could do nothing but accept their decision. The students did not withdraw and waited to see what might happen.

"That night we feared that the Army was going to start a big push at 3:00 A.M., or maybe 4:00 A.M.," recalls one Beijing resident, "but they never did. I stayed on the Square until 5:00 in the morning. By that point it became apparent that the Army wasn't going to move, so we went back home." The warning, it appeared, had been a bluff. Wuerkaixi was forced to resign his position as chairman of the Federation of Autonomous Beijing Student Unions. He does not regret that decision. "When I got the report, I thought maybe it was not correct, but if it was, then what do you do? We hoped it was false, but if it wasn't, we faced a massacre that day. We'd had the same report on May 19. Those who opposed me wanted to sacrifice themselves. The report was false, fortunately, so I had to resign my position."

The protesters began to believe that time was on their side. It was widely assumed that "in Western countries, if martial law forces cannot achieve their objective within forty-eight hours, martial law is automatically cancelled," a Chinese interviewee told me. "People took this to mean that if the government's imposition of martial law fails, the government falls. Of course martial law just continued, but we had this hope." This idea grew out of the commonly held assumption within the Democracy Movement that extraordinary measures in democratic societies have constitutionally imposed limits that are missing in Chinese law. Many supporters of the Movement were thus committed to the idea of holding out until 10:00 A.M. Monday morning, by which time the supposed forty-eight-hour deadline would have passed.

Although the Army did not move Sunday night, the warning may have been more than pure bluff. Some of the detachments around the city had in fact been given the order to prepare to go forward despite the popular opposition, and live ammunition was distributed to some units. But a few hours later, possibly because the students had not been lured out of the Square, the order was rescinded before anything had happened. Had the students ended their occupation of the Square as Wuerkaixi had advocated, the Army might well have forced its way in.

In the end, Monday morning brought what the protesters anticipated and the soldiers longed for: the order for the Army to pull out. It came at 7:00 A.M., forty-five hours after martial law went into effect. Retreat in some cases proved as difficult as advance had been two days earlier. Beijing Television provided a taste of this in a rather frank early-morning interview with the officer of a PLA unit stranded out west in Shijingshan, aired on the news at noon:

"The remarks of Marshals Nie and Xu have just been broadcast," began the television journalist, referring to their message that the students should not fear a crackdown by the PLA. "Can you say something about your view on this matter?"

"I am sorry," replied the officer. "When our unit came here, we did not bring a television set with us. Today we watched the National News Hookup programme, but we missed the part about Marshal Nie."

The officer was asked how his unit had gotten into this situation.

"We could not move forward because we were blocked by masses of people," he replied. "We have reported the situation to the higher authorities. They want us to stay at the present location and await further instructions." The task of waiting had not been an easy one. "Since our arrival here, the masses of the people have not been able to understand us very well. Yesterday morning, many people made a number of provocative and even unfriendly remarks. We told our men to pay them no attention and try to avoid conflicts." This they succeeded in doing.

"Yesterday, our fighters remained on the trucks all day long. They were exposed to the sun and could not eat, but they observed strict discipline. Today, we are supposed to withdraw, but have not done so for one reason or another." He repeated this point later in the interview. "The higher authorities want us to withdraw today. But now we cannot even move back, and the present situation has resulted."

The interview was aired presumably to encourage people to release the troops they had surrounded, to reassure them that the columns they had trapped were drawing back and would not take advantage of their release to double back toward the Square. Given the situation on the streets, such a move would have been impossible anyway.

There were some clashes, most notably down in the Fengtai area late Monday evening. Soldiers there were ordered to withdraw that morning into the nearby August First Film Studio, part of the propaganda arm of the PLA, but the crowds refused to let the unit go. During the afternoon student leaders decided that they would allow the soldiers to withdraw to the film studio so long as they left their trucks and military supplies where they were, but the group army vice-commander in charge of the unit refused that condition. In the end, the unit chose to force its way out of the people's encirclement and head for another destination, a nearby PAP base. A foreign observer was on the scene just before this happened. "The Army column just south of Liuliqiao went back as far as we could see. The line of trucks stretched for more half a mile. The column was locked both from getting into the city and getting out, the back end sealed off with buses. The officers looked disgusted, wearing scowls that suggested they weren't happy with the duty they were pulling. They didn't interact with the crowd beyond saying a word or two. The soldiers had been in their trucks for five days. They were very tired." Some of the troops' fatigue was due to illness as well as exhaustion. A lieutenant in this unit estimated later that by Monday one in five soldiers was sick. When the column at Fengtai began to move, conflict broke out. The protesters misunderstood the soldiers' movements and thought they were starting up to press forward into the city. One of the trucks in the Fengtai barricade was full of bricks. Bricks were thrown. New China News Agency later announced that twenty-nine officers and soldiers suffered mild injuries and were admitted to No. 304 Hospital, a military facility. Roughly the same number of civilians were also hurt, but none seriously.

It was during the Tuesday pullout from Fengtai that the Army sustained its one fatality. A PLA political cadre named Chen Zhiping died early that afternoon from head injuries when he fell off the back of a truck that accelerated suddenly. His death was an accident, and was reported as such on the Tuesday evening news and on Wednesday morning broadcasts of the Beijing International Service.

By the end of Tuesday, the Army's awkward retreat to temporary barracks was basically completed, with a few last remaining units left to disengage the following morning. Student leaders celebrated the rout that day by organizing a protest parade that drew close to a million people, including a hundred PLA naval cadets from Dalian who participated in the march. *People's Daily* reporters also joined the protesters, carrying a banner inscribed "Cancel Military Control, Protect the Capital." Properly speaking, Beijing was not under "military control," as a municipal government spokesman was quick to point out, but the distinction was lost on those who had held back the Army.

Despite their defeat, the troops were not withdrawn from the Beijing region. They were bivouacked in barracks, airfields, shooting ranges, stadiums, and factories, well outside the city but still within one to two hours' range of Tiananmen Square. There was no sign that they were leaving. Logistics centers were set up, suggesting that long-term occupation was anticipated, and security around these areas was tightened. The Army's role in the Democracy Movement was not over.

With its forces sitting and waiting in the outlying areas, the Martial Law Command was obliged to review what it had tried to do, and why its plan had failed. The operation of May 19 fell short of the age-old principle of waging war advised by *The Art of War:*

> Speed is the essence of warfare:
> > take advantage of where your adversary has not reached,
> > make your way by unexpected routes,
> > attack the places your adversary has left unguarded.

The main element in the plan that weekend had been bulk rather than speed or secrecy. Speed was a consideration: trucks enabled the troops to move faster than they could have by foot. So too was secrecy: the Army was dispatched into the city after dusk when it would be less visible. But these were secondary to the main element, which was the intimidation of sheer numbers. This combination of elements had not worked. Partly because martial law was leaked ahead of time, but even more because the citizenry was adamant that the Army not enter the city, the soldiers found no place left unguarded. The offensive had failed within hours of its start.

The lack of prompt contingency plans and close field coordination suggests to me that the Army had not anticipated that its campaign against Beijing would fail. Many units that had to work out alternate routes when planned routes were blocked could not do so because they had no maps to show them other ways to get into the city. Those that did have maps seemed unable to read them accurately. One troop truck carrying thirty-three soldiers from an artillery regiment in Pingshan County on Friday evening to take up a position at Huangcun, well outside the city proper, got lost and ended up at Yongdingmen, the southern bridge into Beijing. To make matters worse, the truck ran out of gas. A crowd of close to ten thousand people surrounded it and pushed it up to the head of the bridge to prevent others from getting through. Like so many other units that night, they could do nothing but sit and wait for further orders, which never came. The Martial Law Command spent the next three days fumbling for solutions without the means to make them work.

The main asset of the operation was simple numbers. The Army hoped that the bulk of uniformed men would impress the people into submitting to military occupation. But was a force of this size really the best means for taking over Tiananmen Square? The question is intriguing because to answer it means that we need to reflect on more than numbers. Rather, we have to imagine the strategy implied by this troop mobilization.

At a minimum, seven or eight divisions (ten to fifteen thousand troops each) were involved in the operation. The total number of soldiers brought into the greater Beijing area was thus in the area of a hundred thousand, although estimates at the time put the number at twice that. Amassing troops on this scale requires considerable coordination over many days. Beijing has several units stationed nearby to protect it, but much of China's military might is located in border areas. Transporting border troops into Beijing is not a simple matter. It cannot be done without several days' notice and several days' more advance planning.

By official report, the elements of this grand army only began to assemble within Beijing Municipality in the two days prior to the offensive. Units from the 28th Group Army, stationed in Datong in the Second Border Region, were not on alert at the time the order went out to come to Beijing. For several nights running prior to May 19, units of the 28th were moved out of Datong by truck along the main highway toward the capital. Its armored vehicles were sent separately by train to the northwestern rail terminus at Qinglongqiao, poised to go in on the night of May 19. But the soldiers of the 28th got no further than the station.

"We found that they had been on the road for several days and had been cut off from the outside world for I don't know how long," reported a Beida student who went out to Qinglongqiao. "They hadn't been allowed to see television or read the newspapers."

A foreign resident of Beijing heard the same stories of several days' travel and almost complete ignorance when he went out to look at the unit from the 63rd Group Army blocked in Fengtai. "They had been in their trucks for five days. They hadn't been told what they were doing nor where they were going. At one point I heard a soldier looking out of a truck and say, 'I think we're in Beijing!' They were told only that they were going on maneuvers."

The problems of transport and logistics meant that the decision to move troops to Beijing on this scale had to have been made at least as early as Monday, May 15, the day Gorbachev arrived in China. If Gorbachev's presence forced the Chinese leadership to eschew strong measures for a week, it also gave them time to bring the military into position for occupying the capital as soon as he had departed. The troops' general ignorance about what they were supposed to do once they got to Beijing seems to indicate that no preparations had been made prior to that mobilization.

Despite almost a week's lead time, some of the units that eventually joined the martial law forces in Beijing were *not* in position by Friday. They were mobilized only after citizens had stymied the first assault. For instance, the vice-commander of a group army from outside the Beijing Military Region says that he received the order to fly two battalions to Beijing only at 7:30 A.M. on Saturday. His unit did not get there until 5:30 that afternoon. (They landed twenty kilometers south of the city at Nanyuan Airfield, a civilian-military facility serving as the major center for airlifting troops and supplies into Beijing during the martial law period.) By the time the officer reached the headquarters of the Beijing Military Region by helicopter to receive further orders, Saturday was over. He and his two battalions remained inactive until June 3. The delay in bringing these battalions to Beijing signals that the Army failed to anticipate that the first wave of soldiers might be inadequate for taking the city. The decisions to call them up and then confine them to barracks suggest that the Martial Law Command kept having to change its plans: first to enlarge the first offensive; then, that having failed, to prepare for a second.

The presence of units like this from group armies outside the Beijing Military Region was not purely a matter of troop strength needs. Another logic was at work. Details about unit identification may make unexciting reading, but from these details emerge telling signs about the thinking behind this troop mobilization. As we shall see, this thinking was guided by political rather than military concerns.

Half of the divisions brought in to impose martial law belonged to the group armies of the Beijing Military Region. One of these, the 27th Group Army, appears to have played a major role in the operation. This group army's responsibility in wartime is to protect state and Party leaders: to come into Beijing, secure the leadership, and get them out to safety. The man in control of the 27th Group Army is Yang Jianhua, nephew of President Yang Shangkun. The 27th is reputed to be tough and to contain more professional recruits than the usual batch of conscripts that make up 70 percent of the PLA. Based in the provincial capital of Shijiazhuang, the 27th Group Army arrived in the Beijing area only on May 19, the day of the operation. University students in the city of Baoding, 120 kilometers south of Beijing, discovered the 27th earlier that day moving up the main highway linking Beijing with Shijiazhuang. Students from Hebei University went out to block the highway, delaying the 27th's arrival in Beijing by several hours.

Other group armies belonging to the Beijing Military Region—the 24th based partly in Chengde to the northeast, the 28th and the 63rd based in Shanxi province to the west, and the 65th based in Inner Mongolia— were also present for martial law duties. The 24th must have distinguished itself in the weeks to come, for its commander, Zhou Yushu, was rewarded by being made head of the PAP in a leadership shuffle in February 1990.

That leaves the 38th, the group army closest to Beijing. The students insisted at the time that the 38th was more loyal to the people of Beijing than to Li Peng and had refused to take part in the suppression. According to an official PLA account published four months after the Massacre, however, the division that spearheaded the invasion of Beijing from the west on Friday evening was none other than the 38th. The presence of the 38th in the operation was also hinted in a brief radio report carried by the Shanghai City Service on Wednesday, May 25. That report says that an armored unit that set out from Baoding on Friday was still blocked well south of the city at Changxindian. Since the 38th has an armored unit, and Baoding is where the 38th is garrisoned, and Changxindian was the temporary base from which Friday's assault was launched, this unit appears to have been from the 38th. At least part of the 38th was following martial law orders.

Movement activists circulated different information. A leaflet written by a PLA officer and printed up by the students reported that General Xu Jingxian, still nursing his broken leg in a Beijing military hospital, refused to lead the 38th into position as part of the martial law force. According to the officer's account, General Xu was visited in hospital on Friday by the vice-chairman of the PLA General Political Department, backed by a squad of military police, to inform him of his dismissal and court-martial.

"Do you have anything to say?" asked the vice-chairman.

"I've already thought this through early on and have prepared myself," he is reported to have replied. "I'm a soldier. When a soldier doesn't follow orders, this is how he should be dealt with. Carry out your orders. I have my own views regarding the student movement, but now is not the time to declare my conclusions." Before being transferred to military prison, General Xu was said to have remarked, "The people's army has a history of never having suppressed the people. I cannot sully that history."

This version of the general's leave-taking is too operatic to be literally credible. But the report in its general outline appears to be true. The main issue for officers like General Xu was not whether the students' views were acceptable, but whether the use of the Army to discipline an urban population was appropriate or professional conduct. Article 29 of the Constitution defines the tasks of the PLA as resisting external aggression and participating in national construction. In the context of dealing with internal security problems, it says only that the PLA must "safeguard the people's peaceful labor." Through the reforms of the past decade, PLA leaders had been trying to downplay the task of national reconstruction and internal security, to reduce the Army's role as a political force or rural reconstruction team and develop it into a professional military corps concerned solely with national security against external threat. To support the Party leadership against student demonstrators not only lay outside the Army's constitutional role but ran counter to the spirit of the reformed PLA. Yet the Party leadership clearly continued to feel that the PLA was a political instrument. It existed to maintain the Party in power and could be called upon to suppress political challenges. Within a Party-dominated Army, there is little room for professionalism.

News of Xu's refusal to command the 38th circulated among the students Friday night and came out in printed form the next day. A leaflet addressed to the beleaguered soldiers informed them, with exaggeration, that "the entire officer corps of the 38th Group Army is opposed to entering the capital to suppress the students." Whatever truth may reside in the students' assertion, they were deceived if they thought the 38th had remained in barracks. At least part of that group army did take part in the first assault on Beijing.

As already noted, group armies under the Beijing Military Region did not make up the entire martial law force. In fact, roughly half the units came from outside the region. Which units were brought in defy complete and precise identification, but educated guesses are not difficult. They appear to have included the 39th, based in Shenyang (headquarters of the Shenyang Military Region); the 47th from Shaanxi province (the Lanzhou Military Region); and at least the 54th from Henan province and probably the 67th from Shandong as well (both from the Jinan Military Region). A unit also came from the Chengdu Military Region, which oversees China's extensive military operations in Tibet; this was probably the 13th Group Army, based in Sichuan province. Units from yet other group armies, like the 16th from the northeast provincial capital Changchun, would fill out the complement of national forces called in for martial law duty at the capital over the next few days.

This national mobilization of troops created for the PLA the problem of how to maintain state security in the areas from which units were drawn. The rapid spread of sympathy actions throughout China in support of the Democracy Movement in the capital made this concern somewhat pressing in the eyes of local military commanders. When the 16th drove out of Changchun on Saturday morning, a foreign scholar who watched the half-hour cavalcade of armed troops parade out of the city asked a Chinese colleague what was afoot. In addition to explaining that these local troops had been called up for martial law duty in Beijing, he told her that units from other provinces were being shifted to Changchun. The advantage to the PLA of bringing in troops from outside the region was that they would be less hesitant to use force to control local supporters of the Movement. According to wall posters in Changchun, the 16th arrived in Beijing on Monday.

Prior to May 22, all but the two military regions of south China— Nanjing and Guangzhou—were represented in the martial law force assembled around Beijing. Their absence may have been simply a matter of distance; it may have been something more. An unidentified source cited in the *Hongkong Standard* on May 22 stated that the commander of the Guangzhou Military Region, Zhang Wannian, had written to Deng Xiaoping opposing the use of the PLA to deal with the student protests. Not only is this letter unconfirmed but Zhang met that day with the administrative and logistics departments of the Guangzhou Military Region to their announce their support for martial law. If Zhang initially resisted the call to support martial law, he had given in by Monday. Individual military districts and group armies within the Nanjing Military Region began issuing statements of support the same day, Monday. One of these was the 31st Group Army, garrisoned in Fuzhou on the southeast coast. Its Party committees announced their support after what the provincial morning news broadcast described as "several days of earnest study"— which suggests that support for martial law was not immediately forthcoming and that some within the local military leadership needed to be convinced. The 31st is one of the three or four southern group armies that were never called into Beijing to take part in the suppression of the Democracy Movement there, either in mid-May or early June. But other units from both the Guangzhou and Nanjing Military Regions would arrive in the next few days to take part in the final suppression. In the end, no military region could refrain from being dragged into the suppression.

The scale of the force amassed around Beijing could signal that the government expected substantial opposition to the troops' entry into Beijing. More to the point, however, the remarkably broad composition of the force suggests that the political leadership felt some uneasiness about the Army's enthusiasm for the task. The Beijing Military Region could have supplied the number of soldiers needed for the operation. There was no logistical need to involve armies from other regions. Their presence had to mean something else, not troop strength requirements but security concerns. The students at the time certainly thought that "soldiers were being moved in from other regions of China, not to put down the students, but to prevent a mutiny by the 38th, as one explained to me later." The presence of an antiaircraft unit seems only to support the speculation that conflict between military units was regarded as within the realm of possibility.

Such conflict could also be seen as mirroring divisions within the Party. It is unlikely, however, that units within the Army were taking sides over the political struggle inside the Party. This was a common suspicion at the time. A satirical poster pasted up all over Beijing during the first week of martial law turned the phraseology of the government's April 26th Editorial against it to pose this suggestion: "A handful of conspirators headed by Li Peng and Yang Shangkun have launched a reactionary coup d'état, compelling Secretary-General Zhao Ziyang to step down. They unashamedly usurped the power of the Party and the state, and moved more than 100,000 Army troops around Beijing. Even now in the 1980s, they dared to impose martial law. This is a fascist coup d'état." Similarly, a letter from an infantry major to the martial law troops that Beida students circulated in leaflet form warned soldiers that they were being manipulated by the Yang Family as part of a plot to allow the "ambitious" Li Peng to usurp the Party and seize power.

The assertion that Li Peng was using the Army to carry out an internal coup d'état succeeds more as rhetoric than fact. Zhao Ziyang was not in a position as of May 19 to protect himself or his faction against Li Peng and Yang Shangkun. He had already lost the crucial support of Deng Xiaoping, and his influence within the CMC had already been curtailed. Li Peng did not need to bring in the Army to defeat him. Having consolidated his power inside the leadership, Li did need the Army, however, to stop the disintegration of his authority from below. The demonstrators calling for Li Peng's resignation had to be silenced, and in a way that would prevent that call from ever being voiced again.

The broad inclusion of units from group armies all over the country indicates that the political leadership had given thought to considerations more sensitive than Li Peng's own power. The Party realized that the order to send the PLA into Beijing was going to be unpopular within the Army. By involving a wide range of units from different parts of the country, it sought to spread the responsibility for suppressing the Democracy Movement so widely that no political or military leader could exploit regional antagonisms and emerge on the strength of regional support to condemn the armies and politicians of Beijing or challenge the martial law order. The entire military structure was now implicated. At the same time, the martial law orders yoked the Army to Li's leadership.

In the event, martial law fell short of Li Peng's hopes. Indeed, it led to the opposite of what he hoped for. The troops got bogged down; their presence breathed new life into the Democracy Movement; and what started as a student action became a broad protest movement. Lao Zi would have smiled at the flowering of unanticipated consequences. He might also have reminded Li Peng that "he who aids his ruler by means of the Way does not use weapons to take the empire." This is precisely what Li Peng had done in the name of Deng Xiaoping. The people of Beijing now regarded Zhao Ziyang as the moral minister who declined to aid his ruler by using the Army; Li Peng's authority was bankrupt.

Li Peng could not have anticipated the citizens' victory, but one wonders why he bowed to it. Why was the Army allowed to sustain this embarrassing defeat? Why was the Army not permitted to salvage its reputation—and Li's—with an order to go ahead at any cost? It seems that political difficulties at the top paralyzed decision making. The Party leadership was incapable of agreeing on what new measures should be taken once the old had failed to produce the outcome it was looking for. The people had reason to be puzzled. As someone at People's University recalled later, "When martial law was declared, we thought they would go in and use force, and when they didn't do it, we were surprised. We began to think that maybe it wasn't only the students that were involved, that there was a struggle at the top that had very little to do with them."

Many rushed to conclude that the government had to fall. As one student observed, in the days following the troops' withdrawal, "The Movement continued to grow daily and the demonstrations never stopped. People began to wonder what had happened to martial law." To another participant, it seemed as though the Movement had won out over Li Peng. "Every day for the next week we heard rumors that the Army was about to enter the city, but it never did, so the students regarded martial law as a complete and unequivocal failure. Once seven days had gone by and the soldiers had not made it into the city, everyone was positive that martial law had failed and that the government was helpless and couldn't bring the situation under control."

The future would reveal this to be a mistaken assumption. The Party may have been momentarily paralyzed, but the government was not about to fall. Nor was the Army about to switch allegiance to the people. It is true that many soldiers were unhappy with martial law duty—and that feeling extended beyond the ranks of the 38th Group Army. But too many students assumed that by fraternizing with the soldiers, they were pulling the lion's fangs. Refusal to carry out firing orders is, of course, what a military command sending troops against civilians most fears, and some regional commanders in the first days of martial law issued circulars to their troops reminding them that they must "strictly observe discipline and maintain their reason. No units or individuals are allowed to partici-pate in, assist, or support the hunger strike activities, petitions, or demon-strations in any form." Some observers felt that fraternization made a definite difference. "People were out persuading the soldiers," one noted, "and these persuasions broke the Army's will. A key factor was that older citizens, their grandparents' age, were often the ones talking to the soldiers and telling them they were making a big mistake."

The will of some soldiers to carry out orders may indeed have been broken by the pleas of citizens, but from that we cannot assume that the Army's will as a whole was broken. If a PLA soldier learns anything in basic training, it is to obey orders. As long as these soldiers had no orders other than to stay put, they gave the appearance of ignoring martial law as much as the people. Under other orders, however, they would act differ-ently. Against the effects of fraternization should be weighed the personal embarrassment and loss of face that many soldiers felt. Few expressed their frustration at being the butt of public scorn and condescension, though we get a hint of it in an interview with two colonels published in the *People's Daily* on May 23, the final day of the pullout. "I think our soldiers have behaved well, despite hunger, thirst, heat and lack of sleep," remarked one of the colonels. "But they feel that they have been greatly humiliated."

Beijing people were too elated with their success at stopping the Army to recognize either this humiliation or the more basic reality that soldiers obey orders. False confidence eroded their natural sense of caution in confronting the Party and its means of violence. As one participant stressed, the students' fear evaporated once it was clear that the Army did not have orders to go ahead at any cost. "Of course we weren't afraid! It was very exciting. We chatted with the PLA, we played cards with them, gave them bread. Our relations with them were very peaceful, very good."

People drew blithe assurance from the belief that right was on their side. Nowhere is it written in the Chinese Constitution that right must always overcome might, nor that the People's Liberation Army will on no account ever fire on the people. But both conditions were assumed to apply. These are assumptions that a regime that relies on military power cannot indulge. They would cost many lives two weeks later, when people went into the streets and were genuinely astonished that the soldiers had been issued live ammunition and the order to use it.

As yet, nothing indicated that the people were wrong to assume that they were stronger than the Army. Nor did appearances suggest that the government was already set on a course that might lead eventually to soldiers opening fire on unarmed civilians. The first truck stopped by the first citizen on Friday night had not pressed forward, and an entire army had been paralyzed. That remained the greatest visible truth, and the greatest false hope. The people might believe, with Lao Zi, that "that which is not of the Way must come to an early end." But they also knew that martial law had not been rescinded, and that the troops had not been sent back to garrison. The passes remained guarded, but the battle was not over yet.

4

Waiting for
the Moon

(May 24–June 3)

By Wednesday, May 24, the Army had withdrawn from Beijing. The PLA had quit the city with its tail between its legs: it looked as though the threat of military occupation was over. But martial law was not rescinded, and the soldiers did not leave the municipality. They encircled it, pulling back to the distant suburbs in two concentric rings, at about one and two hours' distance from the center of the city. The greatest concentration was to the west. The troops set up camp where they could, remained on maximum alert, and waited. No one in the military assumed that the Army's role in settling the challenges of the Democracy Movement was over. No soldier expected that now was the time to go home. There had been no rout. The invasion became an undeclared siege. The Army's inaction did not signal indecision or a vain hope for some nonmilitary solution. The PLA was simply waiting to invade again.

The second assault on the capital would follow a different design. The first offensive had failed because it relied almost exclusively on bulk. The plan had been to propel tens of thousands of troops along multiple approaches into the city and overwhelm the people of Beijing by sheer weight of numbers. That strategy also hoped to exploit the traditional respect the PLA enjoyed among the people. It had included an element of surprise to get soldiers well into the city before opposition coalesced by coming unannounced and after dusk, but otherwise the assault had been above board.

Strategists realized that it would be a mistake to send in large troop convoys from the outskirts once again and expect them to reach Tiananmen Square unhindered. The invader is always at a disadvantage when he must force his way across inhospitable territory to reach his target. This time, therefore, the plan was to have soldiers already in position around the Square *before* the assault started. Then at least some of the troops would not have to first cover a dozen kilometers of hostile crowds. The occupation of Tiananmen Square would not have to depend entirely on the success of the convoys working their way in through citizens' barricades. This would be a double operation. Some troops would come in again along the main access roads, but other troops would already be inside the city.

The new challenge was preparation: how to get large numbers of soldiers past the people's barricades without attracting notice, then move them into position around the Square without anyone's realizing that they were there. The strategists went back to Sun Wu's advice about deception in *The Art of War:* about seeming inactive, about appearing where the enemy does not expect you, about moving moving against him where he is unprepared. They translated these principles into a series of simple, concrete rules of thumb: Disguise the soldiers to look like protesters, not soldiers. Move them individually or in small groups, not in large masses. Dispatch them in the dead of night whenever possible. Send them only in unmarked vehicles, and then only when the citizens' vigilance is down. Where chance of detection is high, have them proceed on foot or bicycle, just like everybody else in the city. Become invisible to your opponent. Then surprise him.

The Martial Law Command realized that infiltrating large numbers of troops into the city required time. The second assault could not happen in a matter of days. It needed at least a week to prepare. We cannot know what determined the final decision to hold the attack until June 3, but one factor deserves mention. It has to do with invisibility.

The first assault took place on the night of May 19 under the light of a full moon. When an American professor told a Chinese colleague that he was impressed with the Army's restraint on that occasion, his cynical friend assured him that restraint had nothing to do with it. "It's a full moon," the old-timer said. "The Chinese army likes to attack when it's dark so no-one can see what they do." The next time, why not wait for the new moon? The darkness would make detection just a little harder, give the element of surprise a bit more edge. The moon would wane to zero illumination at midnight, June 2, and begin to wax in the early hours of June 4. The ideal time, the darkest evening of early summer, would be Saturday, June 3. If this explanation seems too neat, it probably is. The phases of the moon may not have determined the date the Army would move a second time, but it was at the very least convenient to time the operation to the heavens.

Between May 23 and June 3, the PLA had ten days to prepare. The first stage was to assemble sufficient forces on the far outskirts of the city for this two-part operation, and so for the next few days, new detachments poured in by road, rail, and air. Some were units called up before martial law was declared but garrisoned too far away to reach Beijing in time for the first offensive. Among this late-arriving group were seven hundred soldiers from the 39th Group Army, based in the northeastern province of Liaoning. These soldiers showed up in a train at the Shahe Railway Station north of the city on Wednesday, May 24. The station was not secure, however, and they were forced by students and citizens to withdraw farther out.

The troops arriving by air came unopposed. Ten planeloads of soldiers from the Guangzhou and Chengdu military regions landed Thursday night at Nanyuan Airfield, the main gathering point for airlifted troops and supplies south of the city. The soldiers coming in from the Guangzhou Military Region were a special detachment, the elite 15th Airborne Division based in the central city of Wuhan. On maneuvers in central China during May, the 15th Airborne appears to have been the only unit to come to Beijing from the Guangzhou Military Region. Guangzhou's military command did, however, make preparations should it be required to airlift more troops to the capital. It put civil aviation authorities there on notice that, as of Wednesday, May 31, no tickets were to be issued for the next six days. Regularly scheduled flights could be readily canceled on a moment's notice.

The other significant new arrival was a unit from the 12th Group Army. The 12th would be the first, and again the only, participant from its military region, Nanjing. Once the 12th was in place, every one of China's seven military regions was represented in Beijing. By best estimate, then, nine group armies and one airborne division from outside the Beijing Military Region—not counting armored and other specialized units—now had units stationed in the Beijing suburbs. Of the five group armies from within the Beijing region, all were in position, including the wavering 38th. Its loyalty was now assured.

The size of the force waiting on the outskirts was estimated by the student press to be in the hundreds of thousands. Because most of the fourteen group armies each contributed between 7,000 and 10,000 men, the number of soldiers likely exceeded 150,000. This is a conservative estimate, and other observers have speculated that the total was double this number. Given that its opposition was composed of unarmed, untrained, and largely unorganized civilians, the PLA positioned an extraordinarily heavy concentration of soldiers around the city. To the citizens of Beijing, it looked as though their leaders were preparing to use a sledgehammer to kill an ant. It was all out of proportion.

To give the sledgehammer additional weight, the PLA also brought in armor. Armored personnel carriers (APCs) had been present among the forces sent in on May 19, but only in small numbers. Armor had been kept to a minimum on that occasion. By the weekend, though, tanks were detected forty kilometers east of the city. They were in a hold-down position, but nonetheless they were there, out of garrison and waiting. Tanks are not conventionally deployed to handle unarmed civilian opposition. Their role in combat is to counter heavy weaponry and other armor. For citizens, the implication was disturbing. Was there a chance that tanks might be brought into the city?

If their presence was cause for concern, even more so was their eastward location. The armored unit closest to Beijing, the 6th Tank Division, is based northwest of the city. These tanks, however, were to the east. Apparently they had come in from the direction of Tianjin, which is where the 1st Tank Division is garrisoned. The command of the 6th Tank Division is thought to be coordinated with the command of the 38th Group Army, whereas the 1st is affiliated with the 27th. Was the 6th being kept back in garrison for a reason? Had the 1st been called up to impose internal military discipline?

Certainly the PLA did not want even a suggestion of internal opposition to martial law, let alone a hint that there was friction between units. Unity and obedience were essential, and the public impression of unity equally so. By report, President Yang Shangkun on Wednesday convened an enlarged meeting of the Central Military Commission and hammered home to the military leadership its duties with regard to enforcing martial law. The theme of unity was struck the following morning on the front page of the *People's Daily*, which carried statements of support for martial law from the Communist Party committees in the Air Force, the Navy, and six of the seven military regions. Only the Beijing Military Region was not included, though New China News Agency circulated a long article that day about the enthusiasm of the Beijing Military Region's Party Committee for martial law. The committee's formal statement of support for martial law would appear later that day.

The picture of unity within the armed forces was complete, but it papered over considerable opposition. By report, a poster calling for Li Peng to step down was pasted up inside the compound of the headquarters of the Beijing Military Region the same week. Evidently there were still problems within the regional command, possibly linked to General Xu Jingxian's refusal to lead the 38th into battle the previous week.

The May 25 edition of the *People's Daily* also featured a front-page letter to the troops from the PLA Command that struck another necessary theme: rank-and-file obedience to orders. It called on soldiers to "act without hesitation" against what had become a "challenge to socialism." Repeating the essence of the contested April 26th Editorial, it declared that the demonstrators were "plotting to overthrow the Communist Party leadership and the socialist system." Any measure was therefore appropriate to quell this sort of revolt. Although formally directed to the soldiers, the article was a warning to the people that they could not expect the Army to act with vacillation the next time around. And the tone implied that there would indeed be a next time.

The government media also reminded Beijing citizens that PLA troops were encamped in their suburbs. The Army's presence was not disguised. A common theme in official reports was the hardship the soldiers had to endure in their temporary barracks. Newspapers and television featured human-interest stories about local people donating money, food, and other supplies to the units stationed in their areas. These donors were alleged to have come from all walks of life, including independent businessmen, who were widely known to be an important source of funding for the student movement. In almost all cases, significantly, these individuals were left unnamed. The media also reported that some local governments in the outlying areas had formed work groups to assist in improving the soldiers' accommodations. The county government of Shahe, the railway town north of the city, was featured one night on the television news opening schools and a reception center in the county-run cement factory to billet soldiers. Such stories of concerned officials and citizens coming to the aid of poor soldiers projected a false picture of unity between the people and the Army. The happy-camper image may also have misled some into thinking that the nice soldiers could not possibly be turned against the people. But these reports in the Chinese media did convey one simple fact that could not be ignored: the Army was not on its way home.

In those first days after the PLA withdrawal, however, it seemed to Beijing residents that the Army had been driven out of the political equation between the students and the government. True, the group armies had not returned to garrison, but many assumed that the determination of the people to resist military occupation guaranteed that another assault would make the PLA look as foolish as the last one did. If the PLA was being kept at hand, it would be used only in the event of a sudden shift in the political situation. Its presence was a contingency; its reentry was not inevitable. As the days ticked by without any visible moves by the Army to approach the city, the pessimistic feared that the Army would act once the citizens' vigilance was down. The underground student press liked to warn its readers that, as the *News Herald* put it on May 27, Li Peng would bide his time until the moment was ripe. Then, to use a favorite Chinese political cliché, he would "settle accounts" by sending in the Army.

For the most part, however, the student press spoke of the Army's future role in terms of contingencies; that is, the Party would call in the Army only if the students provoked it. One of these assumed contingencies, discussed in the same edition of the *News Herald*, was the possibility that the National People's Congress (NPC) might revoke Li Peng's martial law order.

The NPC is China's pro forma legislative body that rubber-stamps whatever legislation the State Council, headed by the premier, presents before it. This is not the relationship laid out in the Constitution, which places the State Council and the premier under the supervision of the NPC, but in fact this is how it works. In the first few days after martial law, some were suggesting that members of the NPC use their constitutional powers to block Li Peng's order. Fifty-seven members of the NPC Standing Committee submitted a joint letter on the afternoon of Wednesday, May 24, to the Standing Committee general office demanding an emergency session be held within the next two days "to solve the serious problem at hand through legal and constitutional means." The fifty-seven were twenty-six signatures short of the number constitutionally required (that is, half the Standing Committee members) to force a meeting. Moreover, there was no precedent for taking this sort of action from below. Still, they hoped that the precariousness of the political situation might induce the leadership to accede to their demand.

The possibility of suspending martial law and removing Li Peng from power through the NPC spun the political spotlight around onto its chairman, Wan Li. Wan Li, an amiable politician regarded as sympathetic to Zhao Ziyang's reforms but not a powerholder in his own right, was on an official visit to Canada when martial law was declared. After he met with U.S. President George Bush the following Tuesday, it was announced that he would cut short his visit "for health reasons." Wan was expected to arrive in Beijing in the early hours of Thursday morning. That evening, the students dispatched a delegation to the airport to present him with an open letter requesting that a special session of the NPC be convened to cancel martial law and remove Li Peng for exceeding his constitutional powers. They waited in vain. The special airport staff that processes flights carrying Chinese dignitaries was put on duty that night to meet Wan's plane, but it never arrived. The NPC chairman got only as far as Shanghai. There he was met by Shanghai Party chief Jiang Zemin, who would emerge after the Massacre to become the new secretary-general of the Party. Wan disappeared from public view. Three days later, he issued a statement saying that he fully endorsed the decisions of the State Council and the premier, and would not convene the NPC Standing Committee until its regularly scheduled meeting on June 20. The demonstrations he said were "turbulences" instigated by an "extremely small number of people" aimed at overthrowing the leadership of the Communist Party. Over the weekend, other veteran leaders made identical statements. The higher mandarins would not break ranks.

On Monday evening, Radio Beijing's Mandarin Broadcast reported that Peng Zhen, former mayor of Beijing and Wan Li's predecessor as NPC chairman, called a meeting of NPC vice-chairmen that day. It said he wished to make certain that they clearly understood the government's position that some within the student ranks had failed to follow "the orbit of law." The newscaster reported that those assembled were "thankful" to Peng Zhen for providing them with this information. Present at this meeting were only two of the fifty-seven who had signed the Wednesday petition. One of those not present, Cao Siyuan, director of the reform-oriented Research Institute of Social Development of the Stone Computer Corporation, wanted to organize another signature campaign calling for the full NPC to meet right after the June 20 Standing Committee meeting, but his arrest on May 30 (Cao was one of the first proreform intellectuals to be arrested) prevented the move. The contingency of a challenge from the NPC did not materialize. Quite the opposite: the only constitutional means for challenging the legality of the martial law forces had been closed off.

Most people still presumed that the Army would not necessarily invade Beijing a second time. The gradual waning of the Movement over the following week strengthened the presumption that the Army would not be called in—unless the Movement got going again. Some, like student leader Chai Ling, feared just the opposite: that Li Peng was simply waiting for vigilance to fall so that he could send in the Army to wreak revenge against the students. Both views shared a basic assumption, which was that the Army's remobilization was contingent on the Movement's decline. They simply disagreed over what contingency applied. Student activists felt a desperate need to keep the Movement going, whereas most citizens accepted its demise. One victory was the most they could hope for; and anyway, the state always had a way of winning in the end.

The municipal government sought to take advantage of this difference in perception between students and citizens by mobilizing workers to bring the barricades down as early as Monday, May 22. That day, just as the troops started to withdraw, the municipality ordered large industrial enterprises to organize groups of workers into "inspection teams" charged with reestablishing order in the streets. Their main job was to remove the barricades that had gone up during the weekend. The students, naturally, were suspicious of this call for sanitation. By removing barricades, the workers were not just cleaning up the streets; they were clearing potential paths for the Army to launch another attack. A Beida leaflet two days later explained that the workers were being forced to do this work, for which they were being paid a bonus of twenty *yuan* a day, plus new sun hats and free soft drinks. It also reported that some enterprises were resisting the order by complaining of a shortage of personnel. Another Beida leaflet, an open letter from the students to the workers, reminded them that they had been ignominiously manipulated during the Cultural Revolution to suppress the Red Guards in just this fashion, and that they should courageously stand on the students' side in this struggle.

Most workers were not won over to the government's position and continued, in various ways, to show support for the students. On Thursday a small core took the giant step of forming the Beijing Workers' Autonomous Union. Their model was Solidarity. Student leaflets during this week constantly reminded Beijing workers of the heroism of Polish workers, and of the results their heroism had brought about. Lacking the wide base of Solidarity, however, this union was a much narrower organization than the student federation it paralleled and could not extend itself as broadly over Beijing workers as the federation had over Beijing students. Like the student federation, it never gained legal recognition and was actively suppressed after June 4. Though its existence was short, the mere appearance of such an organization was powerfully symbolic for Party leaders, whose Marxism trained them to fear most the mobilization of the working class. Some have since speculated that the Chinese government would never have resorted to force to stop the Democracy Movement had it not been for the involvement of the working class. This is possible, though it should be remembered that the armies were already in position before the union was founded.

The autonomous union was a direct challenge to the official All-China Federation of Trade Unions. That organization existed to carry out the behests of the government on labor issues. Even so, elements within it were sufficiently sympathetic to the demands being framed by the Democracy Movement that the federation on May 18 donated 100,000 *yuan* to the students. Rumors began to circulate during the weekend martial law was declared that the All-China Federation was organizing a nationwide strike to protest martial law. These rumors were so strong that they prompted a federation spokesman to offer an explicit denial on Monday. The denial hints that something of this sort may have been afoot within the government-sponsored trade union.

If any group of workers was on the edge of strike action to oppose martial law, it was the workers at the Capital Iron and Steel Works, an enormous state enterprise situated near the large PLA encampments west of the city. Starting in the second week of May, a small number of activists leafleted workers when they came off shift at many of the large factories around Beijing, including Capital. Workers gave them small amounts of cash to help the Movement, and in some state factories they adopted informal slowdowns to express their dissatisfaction with the government. Capital steelworkers were more active. They organized themselves to join the student-support parades in the week before the declaration of martial law. For Capital to go out on strike would have been extremely distressing to the government because Capital was regarded as a bastion of the kind of working-class conservatism that the Party knew how to manipulate. Action was afoot at Capital the day martial law was declared, for some of the first workers to be tried after June 4 were charged with "inciting" steelworkers that day. Leaders from the central government, possibly including Li Peng himself, went out to Capital that weekend to make sure the steelworkers did not defect to the Movement, according to one of the activists who leafleted Capital.

A visible symbol of the support younger workers were extending to the students in the week following the Army's pullout were the self-styled Dare-to-Die Squads. These bands of eager volunteers anticipated that the Army might come back any time. Their job was defend the students "to the death." Unarmed and committed to the use of nonviolent tactics, they did as much to keep popular support for the students alive as they did to maintain the city's defenses. A Canadian businessman encountered his first Dare-to-Die Squad on Sunday night at Xinjiekou, a main intersection inside the northwest corner of the 2nd Ring Road. They pulled up in a truck and announced that the Army was on its way that night. "They were young, the eldest at most in his early thirties, dressed in workers' clothes with white bandanas around their heads. There was a big crowd there, close to a thousand, and the people listened to them. When they were finished, they drove on, looking for action elsewhere."

The squad's report of an impending invasion was false. Such reports were common fare during these days of lull. On the campuses, student activists similarly tried to keep the notion of the Army's imminent return alive. Yet night after night, the Army failed to appear. People began to doubt that the soldiers would return. As June approached, fewer and fewer citizens manned the barricades. Many came down.

During these days when nothing seemed to be happening, the students debated among themselves the options facing the government. In a published diary entry for June 1, one student wrote that activists at a meeting that afternoon went over what they regarded as the government's four options: to capitulate to student demands; to withdraw the Army and let things return to normal; to let the Army enter the city and occupy certain key points without giving it overall control; and finally, to order a full military occupation of the city. The students regarded the fourth option as "the worst for both sides," and too politically suicidal for the regime to contemplate. "Even for the toughest leaders, those who have lost all sense of being Communists, it would not be easy to court world-wide condemnation in this way and to forfeit the PLA's reputation as the 'people's army.'" They concluded that "the possibility of bloodshed is small, because the price would be too high to pay. Deng wouldn't ruin the achievements of ten years of reform and thereby bring about his own downfall. And in the end, everyone who holds this regime dear would realize that once the army shoots the people, the nature of the whole system changes and the question becomes indeed one of 'overthrow' rather than 'reform.'"

This analysis, though reasonable, was flawed. It failed to factor in Deng Xiaoping's commitment to Party hegemony. Deng may have been a reformer, but he was also a Leninist, believing that the socialist state would collapse into chaos without a strong Party in control. From a Leninist point of view, violence was an appropriate means for ensuring the survival of the revolution—the survival, that is, of the Party. But the reform generation of students forgot Deng's political roots. They thought that Deng's desire to protect his reform program at the same time protected them from violent reprisal.

Regardless of expectations about the use of force, realists within the Movement had to acknowledge that it was losing momentum. It was time to bring the Democracy Movement to an end. On Saturday, May 27, Beida student leader Wang Dan announced at a press conference that the Beijing Federation of Autonomous Student Unions would vacate the Square. Despite strenuous objections from other student leaders, the federation agreed to hold a closing demonstration on Tuesday, May 30. After it was over, the students would withdraw from Tiananmen. There was, however, to be one final gesture. To maintain pressure on the government after the evacuation, they would unveil at that demonstration a permanent monumental sculpture. The sculpture would attempt to express the ideals for which the Movement had struggled; it would also remind the government that Tiananmen Square, even after the students had left it, was still the people's space.

The federation gave students at the Central Academy of Fine Arts, located just two kilometers northeast of the Square, funds and three days to complete the work. The original notion was something along the lines of the Statue of Liberty, an image used in parades in Shanghai during student demonstrations two years earlier, but the sculptors wanted to create something less openly pro-American. They needed to come up with a design immediately. Their expedient, if imperfect, solution was to adapt a studio practice work on hand, a half-meter-high clay sculpture of a man leaning on a pole that he was grasping with both hands. On the assumption that universal ideals take female form, they transformed the man into a woman; his pole was turned into a torch to symbolize the light of truth; and his stature was inflated to monumental size. He was now a woman holding aloft the beacon of freedom. To permit mobility, the statue was carved in four styrofoam sections light enough to be carried. The PSB warned that any truck driver who transported the statue would lose his license, so the students moved the pieces at dusk Monday on a caravan of pedicabs. Anticipating interference, the art students leaked their planned route to the police, then took other streets and arrived unopposed at the Square.

"The statue was to be set up just across the broad avenue from the huge portrait of Mao Zedong, so that it would confront the Great Leader face to face," a graduate of the Central Academy recalled. "When we arrived around 10:30 at night, a huge crowd, perhaps fifty thousand people, had gathered around the tall scaffolding of iron poles that had already been erected to support the statue. The parts were placed one on top of another, attached to this frame; plaster was poured into the hollow core, locking it onto a vertical iron pole which extended from the ground up the center to hold it upright. The exposed iron supports were then cut away, leaving the statue free-standing. The statue was deliberately made so that once assembled it could not be taken apart again, but would have to be destroyed all at once."

Erecting the Goddess of Democracy was a brilliant gesture. Two thousand had showed up for a demonstration in the Square on Sunday; a hundred thousand came for the unveiling on Tuesday. Brilliant, but problematic. Although the statue met the federation's need to draw the people down to the Square on Tuesday for its final demonstration, many regarded the choice of image as a tactical error. Its Statue-of-Liberty style not only insulted the regime but disturbed conservatives who doubted the students' loyalty and sincerity. The government took full advantage of this ambiguity. As the goddess took her place at the north end of the Square, eye to eye with Chairman Mao, the loudspeakers announced: "Your movement is bound to fail. It is foreign. This is China, not the United States." A Hong Kong newspaper reported that the government got the management of Capital Iron and Steel Works to organize a group of steelworkers to go to the Square and knock the statue down, to show the students that they did not approve of this foreign-seeming thing. But the plan was canceled for fear that it would backfire and only increase worker support for the Movement.

The statue had been intended as a parting gesture. Instead, it became a rallying point, especially for a new constituency: students from universities outside Beijing who were arriving in the capital in large numbers. They were not keen to see the occupation of Tiananmen Square come to an end. The federation's decision to vacate the Square had been tied to a resolution urging out-of-town students to leave Beijing. They should go back to their home areas, set up their own autonomous federations there, and build a nationwide basis for the Movement. The newly arrived students did not feel bound by the federation's strategy. It was now their turn to stand in the limelight and voice the call for democracy. The federation bowed to the vote of the students from outside Beijing and allowed the occupation of Tiananmen Square to continue.

With hindsight, many have judged the Goddess of Democracy to be what tipped the scales against a peaceful resolution of the struggle between the students and the government: that this symbolic affront to the regime was too great to ignore, that the consequent revitalization of the Movement had to be blocked. A Chinese participant expressed this view to me as he recalled the split in the student movement Monday evening. "Most Beijing students, including the Students' Federation, felt that major goals had been attained and that it was now time for pursuing a long-term strategy. But young students who had come in from outside Beijing voted down the motion to evacuate the Square. That same night the Goddess of Democracy went up, a serious loss of face for the Li Peng government. Both moves were grave errors."

This view assumes, once again, that the military suppression of the

Democracy Movement was contingent on what the students did during the ten days between the PLA's assaults. In fact, as I shall show, the Army's movements between May 24 and June 3 suggest just the opposite: that the Army was to go into action regardless of student strategies, or worker activism, or the machinations of NPC members. The headquarters of one of the group armies bivouacked in the western suburbs did send out feelers to some of the student leaders, including Wang Dan, to see whether they might be encouraged to give up the Square and return to school. But even had the students vacated the Square, the Army would still have entered Beijing a second time to assert the Party's monopoly over the symbolic power of Tiananmen. The difference would have been only in the level of casualties.

Orders to deploy soldiers inside the city in preparation for the second assault were coming down at least as early as Friday, May 26. I know of two sets of orders issued to military units that day, and I presume there were many others. One set of orders was to PLA video cameramen in Beijing, who were told to get their equipment ready and be prepared to be called at any time to go out and film. The other set of orders went to a unit of soldiers bivouacked in a factory south of the city, ordering them to infiltrate their way into Beijing and take up a position inside the Military Museum by noon on Sunday. The Military Museum, on the far westward extension of Changan Boulevard between the 2nd and 3rd Ring Roads, was closed to the public and serving as a military base for the PLA. It was nine kilometers from where the unit was stationed. The unit's command decided to send its soldiers into the city in single unmarked trucks. Starting at 3:00 Saturday afternoon, the trucks set off at a rate of one every ten minutes. Soldiers dressed up like street toughs, complete with sunglasses, were posted at five intersections along the trucks' route to keep an eye open for potential hazards and signal to the drivers should they have to abort the mission. The unit did not have enough trucks to carry all the soldiers, so the remainder were organized into groups of three to five and sent on foot in street clothes. Citizens did not detect the movement, and all trucks and men were safely inside the Military Museum by 5:30 P.M.

While this unit was trickling into Beijing, another unit to the north of the city was using a different technique to infiltrate its soldiers. This was to leave them in their uniforms, but give them the appearance of being out for jogging exercise. A foreigner who happened to be at the Forbidden City that Saturday noticed a group of five soldiers with a leader, presumably their NCO, jogging southward down through the first courtyard of the Forbidden City toward the gate of Tiananmen. "They were in uniform but not armed. There were people around, but they went unremarked. I assumed they were entering Zhongnanhai to the west or the Great Hall of the People to the southwest, though I didn't see where they went. This was the only such group I saw, though I heard later from other people that the Army had been doing this for days."

These were two of a great number of Trojan horse maneuvers to infiltrate soldiers into the downtown core preparatory to the second assault. The Army began carrying out such moves on Friday and would continue to do so until June 3. Although, as the foreigner in the Forbidden City just testified, people were vaguely aware of some of these movements, no one at the time guessed their scale. Information subsequently published by the PLA suggests the scale was enormous.

It was relatively easy to find secure locations for the great volume of troops being infiltrated into the city, for government ministries controlled many large compounds in the downtown core. Through Chinese media contacts with foreign journalists, we have learned, for instance, that the major media compounds, like *People's Daily* on the east side of the city, the Telegraph Building just west of Zhongnanhai, and the Central Broadcasting Building further west, each received hundreds of troops. The PLA's presence in these places did double duty: the soldiers not only could remain hidden inside the city out of the public eye, but could serve to remind journalists, who had been too eager to join the Movement's call for freedom of the press, who was in control and what was expected of them.

"The staff was aware of their presence," noted a foreign journalist who visited one of these locations, "but they weren't talking with them. They'd been ordered not to run the kinds of stories they had been running and were not to make references to student demonstrations. What they did instead was to write a story straight, and then add in details of what was going on in the city as background, which is something they never dared to do before. In a small way, this was their way of fighting back." Covering the Democracy Movement had touched journalists in a profound way. Once they'd tasted a bit of freedom, they wanted to continue to express what they were seeing in front of their eyes. They didn't want to just pass the Square and pretend it wasn't there. When they got something from the vice-minister's office or from New China News Agency that they were supposed to write, they would add another paragraph or two to express the opposite view. Even conservative Party members of many years were doing it." If the soldiers' presence in media compounds wasn't having the inhibiting effect the government might have hoped for, it was perhaps because the other reason they were there—to be in position for future operations—was more important. The intimidation effect was secondary.

The main locations at which the Army wanted to position large numbers of troops, however, were the buildings adjacent to Tiananmen Square. It has been reported that as early as Friday, May 26, soldiers were being moved through the underground tunnel network into the Great Hall of the People on the west side of the Square, the History Museum on the east side, and the Ministry of State Security just behind that, as well as to parts of the Forbidden City to the north of the Square. The May 28 edition of the *South China Morning Post* reported also that about seven hundred soldiers were camped inside Sun Yatsen Park, which is sandwiched between the Forbidden City and Zhongnanhai at the northwest corner of the Square (see Map 2, p. 132). Tiananmen was thus being surrounded on three sides a week before before the next operation began—and there were soldiers billeted in the Railway Station who could be moved into position at the south end of the Square when the time came.

Hand in hand with the infiltration of troops into the city went reconnaissance by undercover agents. The Army needed to know how things were on the streets: the locations of roadblocks, the resources of the citizens opposing their entry, the general mood of the students and their supporters. To gather this intelligence, every unit dispatched plainclothes soldiers into the city to circulate, observe, and sometimes intervene. As early as Sunday, May 28, the *South China Morning Post* reported that plainclothes security personnel, distinguishable by their green military pants, were seen strolling around the Square in pairs on Saturday morning. An undercover agent working with three out-of-town students attempted to kidnap student leaders Chai Ling and Feng Congde at 4:00 A.M. Tuesday but was foiled when Chai and Feng resisted noisily. Other agents managed surreptitiously to cut the students' telephone and loudspeaker wires.

On Wednesday May 31, commanding officers of units stationed at Nanyuan Airfield attended a briefing meeting where they were told the plan for entering the city and clearing the Square. The target date was Saturday, June 3, the night of the new moon. Presumably other such meetings were held that day in other locations. The next day, June 1, soldiers were moved into final positions on the outskirts—as we know from the cancellation of the 8:00 A.M. tourist train to the Great Wall to allow passage of a large troop train north of the city. Almost all the pieces were in place.

The swelling of the military presence in and around the city during the ten-day lull at the end of May and the beginning of June was accompanied by a growing propaganda campaign to warn people against resisting the Army a second time. The evening news on television repeatedly warned the people about the restrictions imposed on them by martial law, as a foreign student recalls: "There were constant announcements on all channels telling foreigners and journalists not to participate, not to go into the streets, and not to take pictures. We were told that the soldiers had the power to use whatever force was necessary to stop people from gathering in the streets. They wanted to be sure that everybody knew this. They repeated it often so as to weaken the people's determination to resist." The sugar coating took the unique form of popular American films, which were being offered night after night to television viewers between the public warnings. "Why go out on the streets when they've got something better for you?" the student said. "They're not stupid!"

The Chinese media were also trying to manipulate popular perceptions about the Army, to portray the soldiers who had been stranded in their trucks on the weekend of May 20–21 as victims rather than villains. One of the red herrings the government dropped to regain the moral high ground from the students was a retelling of the story about the junior officer who died during the Army's withdrawal on the afternoon of May 23. The national radio news hookup from Central Broadcasting on the evening of Monday, May 29, declared that Political Instructor Chen Zhiping had not died from an accidental fall. Rather, while he was trying to disperse a crowd blocking a column of Army vehicles, someone in the crowd pushed him under a moving truck. A Beida professor who heard the broadcast that evening called the station to ask the news editor how it was that a stopped truck could run over a soldier.

"That's hard to say," the editor replied. He assured the professor that the story was true and that it had come to them from the Army. The professor also called Beijing Television to confirm the original story and was told that this discrepancy would be "reported to higher authorities."

The substance of these telephone calls was printed in a Beida student leaflet later that evening. To check the story, the students were able to find an eyewitness who had been on the scene. He confirmed the original story that the officer's death had been an unfortunate accident, not the work of mob violence. This retelling presaged the propaganda strategy the Army would take a few days hence: to argue that it was not the Army that was initiating the violence against the people but the people against the Army. The government's propaganda had little effect. As one annoyed local resident put it, "The students are the only ones speaking the truth these days."

Rather than trust the people passively to absorb media propaganda, the municipal government mounted a series of public demonstrations in support of martial law during the three days before the Army's reentry. It needed to do this not only to give the appearance of popular support for the Party but to cue the people as to how they should react when the Army came a second time. On Wednesday, May 31, and Thursday, June 1, government organizers staged a round of demonstrations to express "popular" opposition to a "bourgeois" movement out in the outlying towns of Shunyi, Daxing, Miyun, Huairou, and Changping. (On Thursday, the students organized a counterdemonstration in Changping outside the gates of the stadium when local officials were conducting their progovernment rally.) All these locations were well outside the city, where the government hoped the students' message had not penetrated and the local people could perhaps be cajoled into siding with the authorities.

The rally in Daxing ended with a bizarre and disturbing ritual. Party officials burned two men in effigy, one of them astrophysicist Fang Lizhi. Targeting Fang was a surprise to everyone, especially to the peasants at the rally, who had never heard of him. Like the Soviet scientist and dissident Andrei Sakharov with whom his name is often linked, Fang became China's most prominent human rights advocate out of the simple conviction that without an open environment of ideas, science could not flourish.

The struggle to bring about that openness led him in the 1980s to voice broader pleas for greater democracy in China's political and educational systems, until the student movement that unseated Hu Yaobang forced him to withdraw from the public realm. "I didn't take part in any of the students' organized activities in the spring of 1989," Fang told me later. "But the government seemed to feel that my ideas had been a factor stimulating the Movement." Not only had Fang remained out of the Democracy Movement but he had criticized his students for going too far, particularly with the hunger strike. "Once the hunger strike started, the Movement went out of control, and I suspected that the government would use military means to end it. These students just did not understand. They grew up in the generation after the Cultural Revolution and had never seen the Party kill people on a large scale. The students loved that line in *The Internationale* about this being the final struggle, but I told those who came to my home that this was most definitely not the final struggle." Fang shook his head sadly. "They felt that if they just carried this struggle through, they would be victorious. I didn't think so."

Fang Lizhi's criticial view of the students' strategy meant nothing to the Party. Nor did it care that Fang had actually been out of Beijing at an academic conference for the week leading up to the Daxing rally. Party propagandists needed a convenient public figure to tar, someone who could be identified as a counterrevolutionary "black hand" behind the movement, manipulating innocent students. When NPC Chairman Wan Li two days earlier had declared the protests to be the work of "a small minority," he was mouthing the government's face-saving formula for crushing the Democracy Movement: arrest a small number of intellectuals who over the past few years had annoyed the Party with the call for a more democratic system, and the government could be seen as punishing the protest without having to put the entire student body of Beijing into prison—which the government well recognized it could not do. Fang would be the scapegoat, regardless of what he did or did not do.

City workers were less easily cajoled into attending government rallies than their rural counterparts. During the first two days of June, the government demanded that urban enterprises and government offices stage a second round of demonstrations within the city in support of martial law. The results were laughable. When the staff at the prestigious Academy of Sciences was asked on Wednesday to attend a demonstration in support of Li Peng the following day, only one person volunteered to go. In another work unit, office workers were given a cash payment of ten *yuan*, a theater pass, and half a day off work to attend a demonstration that Thursday. One of them took his payment down to the Square and gave it to the students. Workers at the New China Printing Press were offered twice that fee plus a free strawhat to join a pro-martial-law demonstration on Friday: not a soul turned out.

The demonstration that did capture public interest on Friday was on the Square. Four prominent individuals decided that afternoon to stage a final hunger strike on the steps of the Martyrs' Monument. This was intended as a last farewell gesture of defiance before the Movement slipped away. The strike would involve only the four of them and last for only seventy-two hours. Then they would depart. The leading figure was Hou Dejian, a popular Taiwanese-born singer/songwriter who defected to China in 1983. He was joined by Liu Xiaobo, a literary critic who had just returned from a stint at Columbia University; Zhou Duo, a sociology lecturer at Beida; and Gao Xin, former editor of the Beishida campus newspaper.

The strike manifesto they issued rejected the notion that the Li Peng regime was the "enemy." It asked only for cooperation from the regime in creating a civil society of "legal autonomous organizations" that would create a democratic balance to state power. Within an hour of the announcement, the Square came back to life. "When word spread that Hou Dejian was leading singsongs, people started to pour back into the Square," recalled a foreign journalist. "It was festive, and sagging hopes were no longer sagging. People's spirits picked up." She left the Square later that afternoon, but the momentum continued to build through the early evening. "When I got back about 8:00 that night, the difference with the afternoon was like night and day. People were enthusiastic and excited. They were singing. It was like blood flowing through a vein."

As these new developments animated the Square, several young demonstrators forced a car without license plates passing Tiananmen to stop. They were suspicious that this was an Army car, and that the driver had removed its military plates so as to be able to drive through the city without attracting notice. They were right. Inside rode several PLA officers, one with the rank of major. A fight almost ensued, until students intervened and took the officers to the Monument for questioning. All they could learn was that there would be troop movements that night.

In fact, there had been movements of PLA troops into Beijing all day long Friday, just as there had been for the week previously. But the volume of troops being infiltrated into the city on that day greatly exceeded the number that had been brought over the previous week. Most of them, it seems, were going to the Great Hall of the People, the PLA's largest base inside downtown Beijing. The Great Hall was being prepared as the command post for field operations in the upcoming operation to take the Square. Divisional commanders would be driven in after midnight to convene an organizational meeting at 3:10 in the morning.

The largest known infiltration of troops to the Great Hall on Friday was an entire division, over ten thousand men, from what I guess to be the 65th Group Army. The published PLA source that tells of this infiltration says only that the troops got to the Great Hall in plainclothes and does not reveal how such an immense movement was disguised. The same source gives more details regarding another infiltration, code-named F75, a reference to the seventy-five hours soldiers had sat besieged during the first offensive. This unit divided up into more than twenty smaller groups and moved in plainclothes through the streets of Beijing Friday afternoon. To guarantee authenticity, some of the unit went out to clothing stores Friday morning to buy suitable outfits: the plain cotton trousers of the worker, the sporty jacket of the tourist, the printed T-shirt of the street youth. What they found they couldn't disguise were their army haircuts and the marks that years of wearing military capbands had printed into their foreheads. A few were stopped and questioned about these details, but they were all able to parry the challenges of suspicious citizens and reach their destination by 6:00 P.M.

Infiltrations by large numbers of plainclothes soldiers continued Friday evening. The soldiers of a regiment bivouacked down in Fengtai in the southwest suburbs had to cover a dozen kilometers to reach the Great Hall. It was decided that they should move on foot in small contingents of about a hundred at a time, well spaced through the evening so as not to arouse suspicion. The first contingent was able to make the trip unopposed, arriving at the Great Hall at 7:30 P.M.

Yet other units, who knows how many, came in from other directions, as alert individuals began to note. Shortly after 8:00 P.M., a foreign aid officer spotted two Army trucks parked north of East Changan Boulevard between the 2nd and 3rd Ring roads. A student command vehicle normally stationed there every night had not shown up since Wednesday, so the trucks were able to drive in undetected. "The trucks were empty. There was no-one in the back, and I didn't see anybody in the front. We later learned that the troops in these trucks had infiltrated across the 2nd Ring Road and gone in through the alleys to get downtown. We had a sense of foreboding when we saw them, because we sensed something had to happen." Someone else noticed another small group of military trucks about 10:00 P.M. parked near the Beijing Zoo near the northwest corner of the 2nd Ring Road. Again, there were no soldiers, though the observer did notice boxes of supplies in the back.

These trucks were small parts of the final night of infiltrations preparatory to the second assault on Beijing. The most conspicuous part Friday night was a column of some eight military vehicles followed by several thousand infantry who came in from the west along Changan Boulevard. According to one estimate, these soldiers were from Chengde, members of the 24th Group Army. They were dropped from trucks just west of the 3rd Ring Road about 10:30 P.M. and then proceeded east on foot. At 10:50 P.M., secrecy was shattered. A large Mitsubishi jeep belonging to the People's Armed Police overtook the head of this column at Muxidi. There are conflicting accounts as to whether the vehicle was traveling from the west with the column, coming down on it separately from the north, or simply driving out of the nearby compound of China Central Television (CCTV), but everyone agrees that it was moving at a dangerously high speed. Street-cleaning trucks had been through shortly before and the surface of the road was slick. The driver lost control of his vehicle by the overpass and veered off into the bicycle lane near the entrance to the Muxidi subway station. It crashed into three cyclists—a woman, a child, and an elderly man—and hit a pedestrian before coming to rest on the sidewalk. The injured were rushed by ambulance to Fuxing Hospital a few blocks away. One died en route. Two more died in hospital.

Bystanders converged on the jeep in an angry crowd. There was a driver in the front and three soldiers in the back, none of them in uniform. A foreign student who arrived on the scene within ten minutes of the accident saw that "the bikes and bodies were all twisted up. The crowd was really angry. The vehicle was smashed by the crowd just after it happened, but the occupants inside were unhurt." When the driver said he refused to take responsibility for the injuries because he was under martial law orders, young men in the crowd pulled him out of the jeep and slapped him in the face. The soldiers in the back leapt to his rescue and then they raced off. A PSB cruiser was on the scene within a couple of minutes and, according to one student source, spirited two of the soldiers away. An hour later a PSB tow truck arrived to drag the demolished vehicle away, but the crowds blocked its passage and made it impossible for the police to remove what people regarded as evidence of the Army's callous treatment of the citizens of Beijing. "We were safe before the Army entered the city," declared one spontaneous orator. "Only now that the Army has come in are our lives in danger."

The official account of the accident on the television news the following morning declared that although the jeep did belong to the PAP, it was on loan to CCTV and not involved in any military operation that night. Students reported that they found an envelope in the vehicle from the Martial Law Command addressed to CCTV marked "Urgent and Top Secret," also that the jeep bore no license plate. Because police and military vehicles carry distinctive license plates, it is possible that the plate was removed when the vehicle was loaned to CCTV, if indeed it was, to make clear that it was not being used by security forces. More likely it was removed to ensure that the jeep would not be spotted and stopped as a security vehicle. A foreign traveler watching the news report on the accident in a hotel lobby next morning recalls finding the official explanation ludicrous. "I remember this clearly, because even though a lot of the people who were sitting around watching had never been to China before, everyone recognized it was a big lie. The newscaster said it in such a way that it was so obvious, and everybody in the room started laughing when he said it. It was apparent that they were pretending it wasn't the Army that had done this."

The news of the accident spread like wildfire throughout the city. The alarm was up. Everyone assumed that the jeep was a harbinger of a second military invasion, "an important signal that the Army might be about to move," as a Chinese graduate student told me later. Indeed, an onlooker warned a foreign eyewitness at the scene of the accident, "Don't stay here. Something nasty is going to happen tonight. Go home." In fact, he was wrong. The Army would try to avoid confrontations Friday night. By the time police cars with flashing lights arrived on the scene, the Army had pulled back the infantry column coming in from the west. Its presence had gone undetected. Elsewhere, however, other infiltrating units were being discovered and hindered. The regiment from Fengtai that had been able to slip its first group of soldiers into the downtown core earlier in the evening ran into difficulties with the rest of them. Over two thousand soldiers ended up being divided and blocked by crowds along Xuanwumen Avenue and Changan Boulevard. Some smaller groups managed to sneak off through back alleys and get through to the Great Hall, but by dawn half the unit was still stranded in the west end of the city.

A foreign journalist found them along West Changan as far out as Muxidi several hours after the jeep accident. They were dressed in white shirts and green military pants but without other sign of uniform. They seemed disoriented and to be without commanding officers. "Groups of people ran along beside them, grabbed them, not trying to hurt them," recalled the journalist who was there, "but they were excited and were interrogating the soldiers. Some of crowd roughed the soldiers up, even ripped their clothes; some of the soldiers were crying. They seemed to have no idea what was going on. They didn't have any weapons, though some of them had two belts, as everyone noted." These soldiers were under strict orders not to strike civilians, even not to use their belts, as one of them told student leader Wuerkaixi when he went out to inspect the situation. "Don't worry," the soldier said. "I have been ordered not to lay a hand on anyone. We have all been ordered not to lay a hand on anyone." These orders covered every operation throughout the city Friday night: all conflict should be avoided. This was not an attack.

The next major infiltration, the boldest of the night, came three hours later. The six thousand soldiers whose approach from the west had been aborted by the jeep accident would try again on foot, but from the opposite direction. They were moved by truck around to the east side of Beijing, somewhere just beyond the 3rd Ring Road, then they jogged in by battalion along the eastern extension of Changan Boulevard at about 2:00 A.M., heading again in the direction of the Square.

"The soldiers were running in short-sleeve white shirts, green pants, and running shoes," according to a military observer who spotted them jogging in formation past the Jianguomenwai diplomatic compound. "They didn't have proper gear on their backs. They had make-do packs, anything, even cardboard boxes. They were young—none was over forty-five years old." Observers at the time guessed that these soldiers had jogged the twenty-plus kilometers in from Tongxian, but this eyewitness did not think so. They were fresh when they passed the 2nd Ring Road.

"There was even an officer carrying a clamshell-style briefcase full of documents. He was holding it by the handle and running with it. You can't run twenty-odd kilometers carrying that. Some of them had radios with antennas and headsets, and again, no sign of stress. Finally, no-one was trailing behind. I went out to Tongxian at 4:00 a.m. and didn't see a single straggler between the 2nd Ring Road and Tongxian."

The jogging soldiers were detected as soon as they crossed the 3rd Ring Road, and a squad of cyclists preceded them by about five minutes in Paul Revere fashion, shouting, "The Army is coming! The Army is coming!" Barricades had not been prepared along Changan Boulevard that evening, so citizens raced out of their houses as soon as the cry went up to assemble makeshift barriers hastily. Trucks were pushed out on the street before the soldiers arrived, but this tactic, so successfully used against motorized columns, was ineffectual against platoons on foot. They simply ran around the trucks and went on. The only effective barricade that night was at the Dongdan intersection east of the Square, but it went up only after the troops had already run past.

When the soldiers reached the intersection just east of the Beijing Hotel, two blocks from the Square, they were met by a mass of people who stopped them more effectively that any truck barricade could. The foreign military observer saw the river of soldiers break up. "Right in front of them was a human wall about a hundred people deep from one side of the street to the other. That's what stopped them. They lost the Square by just a few minutes. If these guys had reached this point barely three minutes earlier, the human wall wouldn't have had time to form." The Army missed by a hair's breadth. Their window of opportunity closed, the soldiers hesitated a moment, lost their forward momentum, and were overwhelmed.

At first the soldiers tried to break free from the crowds, but the people were determined not to let them escape under cover of darkness, and most had to remain where they were. Within fifteen minutes, much of the column had been broken up into clumps of several hundred soldiers scattered along Changan and up a few of the sidestreets, each clump isolated from the other by the citizens who had surrounded them. "When people yelled at them to sit down," a journalist told me, "they all sat down. No officer was there to provide orders. They just sat there. You couldn't tell where they were from, and of course they wouldn't say anything. A few had radios, but they didn't seem to be using them." For the next four or five hours, these six thousand soldiers were marooned islands in a sea of citizens. "The crowd would not let them leave and wanted to wait for dawn so that the entire populace of the city would see the government's cunning," as a student handbill printed Saturday morning described it. "That made the soldiers even more desperate to withdraw. About three hundred students from the Central Nationalities Institute tried to block them, but with one big push, some of the soldiers broke through and took off." Toronto *Globe and Mail* correspondent Jan Wong was heading west on Changan when she encountered this group of soldiers running back east in disarray at Dongdan. "They were running away from the direction of the Square, yelling and trying to get into compounds of different buildings. They weren't in uniform, but they had the shaven heads, white shirts, green pants, and running shoes of Chinese soldiers. Some of them had communication gear on their backs. Some had heavy cables coiled around their bodies." Fleeing soldiers dropped their food rations. Dried noodles littered the streets and crunched under foot.

Most of the soldiers remained trapped back in the vicinity of the Beijing Hotel, as Wong discovered when she got there. "It was dark and these soldiers were scared. They didn't know what was happening. The low-ranking officers with them didn't seem to know what to do either. They didn't know where they were. They were scared and stranded and there was no plan." Others on the scene were also struck by the apparent absence of commanding officers. There were a few junior officers and senior NCOs with these soldiers, in fact, but not many. A military observer noted that "the ratio of officers to soldiers was very small, maybe one officer for every two or three hundred soldiers. And they were young, only as high as captain rank. Also, when the column got stuck in front of the Beijing Hotel, the tail end was almost all the way to Dongdan. Most of their leaders were forward, so the ones at the end had no senior experience whatsoever. A small group of three or four hundred got isolated without senior NCOs or officers."

The officers who were at the front did not know what to do once they had fallen short of their goal. Their indecision gave the impression that this group of soldiers was a rudderless vessel becalmed in a human sea. But there was nothing they could do. It must be remembered that field officers in the Chinese military do not have authority to use their own initiative to deal with unanticipated situations. This sea of people in their path was a contingency the strategists had not reckoned with. The officers were at a loss as to how to extricate their men from this humiliating encounter with city residents. The plan had failed, and there was no backup.

Unfamiliar with the layout of the city and without even a city map, the officers had no idea even where to retreat to. At last they decided on the Railway Station, but even then they had to ask the people for directions. The column had passed the station a few blocks back during their jog toward the Square, but their officers did not know where it was. Gradually, by about 4:00 A.M., they were allowed to make their way, some without their shoes and all in disorder, to the station. There they retreated down into the subway. The crowd closed off the exit behind them. Unable to get out, some of the soldiers walked on through the subway tunnel to the next station at Jianguomen, but the crowd got wind of the movement and blocked that station as well. The stairs leading up from the platform at the Jianguomen station are long and steep, which made it too difficult for the soldiers to push their way past the demonstrators. Several thousand troops thus remained caught between the two subway stations until the following morning. But hundreds more never even made it that far and wandered lost in the maze of alleyways that meander through the old sections of the city north and south of Changan.

A few soldiers ended up on the Square, some by accident and some under duress, as one journalist walking in the Square discovered. "A man came running up to me. He didn't have a jacket on, only a T-shirt, and he didn't have shoes or a hat. He told me that he was a soldier." The journalist was puzzled and doubted the man's military identity. "I remember thinking he didn't look like a soldier. A bit later I saw three flatbed bicycles go by piled with stuff. Two of the flatbeds had people on them, covered in clothes," hidden apparently to prevent people from discovering their identity. "Some students lifted the clothes off and took the people underneath out, treating them like prisoners. They were soldiers. The third flatbed was just clothes—army uniforms, jackets and hats and shoes. They had no more than five prisoners, maybe four, and they marched them up the west stairs of the Monument." These soldiers were returned to the Army later that morning.

The vigilance of the citizenry destroyed the Army's plans to infiltrate troops Friday night. At a moment's notice, people in various states of undress—undershirts, high heels, leopard-skin dressing gowns—filled the streets and tipped the scales against the Army's footsoldiers. This resistance arose not only along Changan Boulevard but elsewhere around the city. When a battalion of infantry tried to penetrate the south end of the city at about the same time, these soldiers too were discovered and blocked. A thousand residents stopped them on Dongjing Street near Friendship Hospital, two kilometers short of the Square. According to a government source, 24 of the 340 soldiers were injured and had to be treated in the hospital. The rest took sanctuary there.

These were only the infantry infiltrations on foot. The PLA also moved troops and equipment that night by bus, with some success at first. But once the citizens were alerted, these attempts too were thwarted. A Canadian journalist learned that something was happening on the other side of the city when, about 4:00 A.M., a woman told him that the Army was teargassing its way into the city in a major assault from the west. "First we went to the Square. Students were moving up to the northwest corner en masse to look down Changan Boulevard, though the street was clear. There were no signs of soldiers. But some of the demonstrators in the Square told us that if we walked west along Changan, we would eventually run into the Army, which had tried to make a move on the Square from that direction." The journalist and his cameraman decided to walk west past Zhongnanhai. At Liubukou, the first major intersection beyond Zhongnanhai, they found huge crowds of people. "A small bus was parked just to the south of the intersection, fifty feet from the corner. It was full of young men dressed in white shirts. The people who had stopped and surrounded them were pointing and laughing at them. They looked embarrassed." On the far side of the intersection, an officer's car sat on flattened tires, surrounded again by civilians. "An automatic rifle sat on the dashboard. The officers inside looked absolutely murderous. One chap in the car was gritting his teeth and staring straight ahead. It was a dreadful humiliation for them to be surrounded by civilians during martial law, unable to move. Other buses further west were also stopped by civilians."

There was no sign of tear gas, and it was hardly a major assault, but the story of soldiers trying to make their way into the city was otherwise true. No one at the time guessed that these soldiers were part of a large motorized infiltration of troops from the west that had started just after 1:00 A.M. An unknown number of buses were involved in the operation, which was to ferry soldiers and supplies on hourly runs to the Great Hall of the People. Each bus carried over thirty soldiers. Because of the jeep incident earlier that evening, citizens were blocking some of the roads into the city and military drivers had to take circuitous routes to escape detection. One bus on its first run needed an hour to cover a distance that should have taken no more than twenty minutes in order to evade roadblocks in the area of the southwest corner of the 2nd Ring Road. As it emerged into the Xidan intersection a little after 2:00 A.M., it was spotted and stopped. Its windshield was smashed, its tires were flattened, and there it would remain for the next twenty-one hours. This was the first of the buses and jeeps that would be immobilized on the west side of Beijing over the next hour and a half. The bus at Liubukou that the Canadian journalist ran into was stopped at 3:00 A.M. on its third run into the city. By 3:30, five thousand people choked the Liubukou intersection and made further traffic along Changan Boulevard impossible.

Dozens of other buses, jeeps, and trucks coming into the city by various routes met the same fate. Some were stopped in the far outskirts. Others were blocked trying to use side streets between the 3rd and 2nd Ring roads. A few others made it closer in to the downtown. One bus full of soldiers was blocked on the street directly north of the Forbidden City. Some of these vehicles were able to turn around and pull back before crowds surrounded them, but most were caught by angry citizens who flattened their tires and posted guards. Neither the buses nor the men inside could leave without the citizens' permission, although one driver whose bus was stopped near Xidan managed to sneak away from his captors on a washroom break later in the day.

It was the clandestine air of the operation that angered citizens. Soldiers were sneaking into the city in semidisguise after the citizens had made it clear that they were not wanted. As a Chinese eyewitness at Liubukou observed, "The soldiers were not in uniform, and the people were enraged at the secret moves in such contrast with the slogans on display: 'Learn from the people of the capital! Salute the people of the capital!'" They not only were in plainclothes but were traveling in non-military vehicles; instead of troop trucks, they came in unmarked buses that bore the red license plates found on public transport, not the white plates that distinguish military vehicles. The citizens felt outraged that the PLA should use such deception against the people, though of course the Army had been doing this for a week by the time it was caught out.

The fire of popular anger flared when people discovered that these vehicles were carrying not just plainclothes soldiers but weapons, and in considerable numbers. At first, officers tried to hide their small arms (one gave his pistol to his driver, who stuck it in his underpants) and the men in the buses kept their guns concealed under kit bags under their seats. The PLA did not intend that the citizens should know about the weapons, nor did the soldiers have authorization to fire them, and they did their best to prevent people from coming onto the buses and investigating. But citizens did succeed in entering, in one case by a roof vent, and found automatic rifles, hand grenades, even a cache of homemade knives. These they brought out and showed to the crowd.

"Some of the students and workers who were surrounding these buses began to pull out guns," observed a Canadian television journalist. "They did not pull all the guns out, only a few, and they showed them out the windows to the crowds, and to the media who were there. Several protesters got up on the roofs of the buses and proudly displayed these weapons, setting up a little shrine with helmets and some guns. But they were told by other people in the crowd to put the weapons away, and eventually the weapons were put back in the vehicles." Some buses carried only the arms assigned to the soldiers on board, but others were being used primarily as weapons transporters, ferrying AK-47s and machine guns in burlap bags under the custody of half a dozen soldiers. One bus was carrying nothing but ammunition. Clearly the Army was endeavoring to build up its supply of combat weapons within the city before the next assault.

Despite Friday night's failures to sneak men and equipment past the citizens, the Army continued its attempts at small-scale infiltration during the day on Saturday. A foreign businessman working in the east end of the city noticed the remains of one such attempt: an army jeep abandoned during the morning in front of the International Club. "On closer examination, you could see that someone had lifted the hood of this jeep and ripped out some wires. It was undrivable, so the soldiers had deserted it. While I was down there in the street having a look, another vehicle came by—a troop truck covered with a tarpaulin. When it slowed because of traffic, it was surrounded. Bodies appeared like dust coming to a magnet." The maneuver was the same as it had been for the jeep. In thirty seconds, some young factory workers had lifted the side flap over the engine and ripped out the wires. The driver, a tough-looking guy in uniform, leapt out of the truck.

"What are you doing?" he yelled in a threatening voice.

"You killed three of our people last night," they yelled back, referring to the jeep accident. "You should get out of our city!"

Their aggressive posture confused the Army driver, and he retreated cautiously back into his cab.

"Okay, maybe three people were killed," he said, "but I don't know anything about that." After some discussion, he agreed to turn his truck around and drive back out of the city. By this point, though, the truck was completely disabled. He couldn't get it going and it remained stalled at the side of the road.

As the foreigner stood watching this scene, another truck drove by in the hope of slipping unnoticed through the traffic, but was stopped in the same way. "The guys in the back of this truck looked like teenagers, maybe seventeen, farmboys who didn't even know how to wear a uniform. They looked so unsoldierly, as though they hadn't been trained to do anything. The few in charge, like the drivers, looked like real soldiers, but the others looked like new recruits. After their trucks were disabled, these soldiers beat a retreat on foot." By this time an aggressive crowd had gathered and was on the lookout for more suspicious vehicles.

"Then a third truck came by. Traffic was moving and it wasn't going to stop. People raced after it, and it wasn't until they were up on the Jianguo-men overpass [a block further west] that the traffic forced the truck to slow down, enough so that three or four young guys in the front who could really sprint got up to the truck. They leapt on it right by the doors while it was going, pounding on the window. One of them maneuvered his body onto the hood before the truck even stopped. The driver braked. That gave them just enough time—the hood went up and the wires came out in a ten-second operation." The young men had the procedure of incapacitating trucks down to a science. They were pleased with their power. "Just before throwing the wires, the young man who cut them held them up to the driver to see that he wouldn't be able to drive away, then threw them. The crowd was emotional and angry, but they could see right away that these soldiers were unarmed and they let them go." A student handbill printed later on Saturday asserted that this sort of infiltration by single vehicle went on elsewhere in the city that morning. Small groups of soldiers in trucks, minibuses, even ambulances, were discovered and stopped, some out on the edge of the city, some within a few blocks of the Square.

The Army on Saturday morning faced a double challenge: to keep feeding soldiers into the downtown, and to recover those who did not reach their destinations. Recovery was not a simple matter. Where the distressed soldiers were few and on foot, the Army could dispatch plain-clothes agents to gather them up and lead them to safety. But where they were many or conspicuously exposed, as the armed soldiers in the buses on Changan Boulevard were, less subtle tactics had to be found. The AK-47s in particular had to be recovered. Facing the choice of seizing them by force or negotiating for their return, the Army chose seizure.

The first attempt failed. A team of thirty soldiers was sent from the Great Hall at 10:00 in the morning, but only seven were able to reach the bus at Liubukou. Then all they could do was negotiate, first with student monitors, and later with the student leaders in Tiananmen. (Not until the evening did the students agree to let the PSB transfer these weapons to the Navy Hospital.) An official Chinese publication declares that one soldier was injured and admitted to the China Emergency Center, a medical facility four blocks south.

The second attempt to recover the weapons came at 2:00 P.M. Government loudspeakers at Liubukou asked the crowd, now grown to tens of thousands, to release the vehicle carrying arms and return any weapons that had been removed. Ten minutes later, the first round of tear gas was fired. The soldiers misjudged the wind and the tear gas blew back in their own faces. The crowd's immediate reaction was anger rather than fear, but it was suddenly met by some three thousand helmeted riot police who rushed out of the main gate of Zhongnanhai swinging nightsticks. The police were greeted with a barrage of paving stones and soft-drink bottles. A second round of tear gas forced the people back, but the seesaw between the crowd and the police continued for almost forty minutes. According to two elderly residents who kept count, ten canisters of tear gas were fired at the crowd. The riot police succeeded in cordoning off a portion of the street and getting to the bus, but they had to retreat north under a hail of rocks and bottles, returning to Zhongnanhai via the west gate.

More than forty protesters were injured, and at least twenty were taken by ambulance to No. 2 Hospital. One was a worker who broke his leg while trying to escape over a wall from the riot police. Another was a student at the Beijing Agricultural Engineering University, who said that he was struck on the side of the head when he tried to stop two plainclothes officers from beating people. The tear-gas canisters, which bore the marking "Metak 38m MN-05," also did some mischief. A few people suffered direct hits, and others were burned or cut by pieces of casing when canisters exploded near them. "I saw a guy who had been hit in the face with a canister," recalled one eyewitness. "His face was covered in blood. He had the canister in his hands and was waving it around."

After this little battle was over, a second squad of about forty soldiers emerged from the main gate of Zhongnanhai and formed a tight defensive line in front of the leaders' compound. They were the object of abuse and petty retaliatory violence, as an eyewitness discovered when she arrived in midafternoon. "People were throwing rocks at the uniformed men. They weren't responding: they didn't use their batons or throw anything back, but the crowd was beginning to take it out on them. All the windows of the military buses parked nearby were broken. There was glass all over the street. People were prying off the license plates, brandishing them as evidence. I also passed three or four sidecars to Army motorcycles that had been disabled, their tires flattened on either side." At 4:40, security officers tried again to disperse the crowd with tear gas, but without success.

Shortly after the first tear-gas attack, the crowds on Changan Boulevard were swollen by the arrival of a large student parade heading from Beida to the Square. A student in the Beida demonstration recalls that, before they reached Xidan, "we saw a man about thirty or so who was crying, presumably from the tear gas. He was carrying a little girl of three or four, who seemed to have fallen asleep. He said, 'I am a Communist Party member. But I tell you not to believe the government again. My little girl has been hit by tear gas.' We told him to rush her to the hospital, but he said that they'd just left the hospital. His face was streaked with tears."

Because the section of Changan where police had battled with demonstrators was now impassable, the column of students had to turn south and proceed along side streets leading to the west side of the Great Hall of the People. There they encountered yet another standoff between citizens and security forces. A thousand uniformed soldiers had emerged from the western entrance of the Great Hall at exactly the same time the riot police burst out of Zhongnanhai under a cloud of tear gas, part of the same operation to recover the equipment on Changan. They were completely blocked. In the words of a student participant, "The street was a sea of humanity. I heard others say that the soldiers had beaten some people, so they had been surrounded. I got in closer to the fence around the Great Hall and could see the traces of blood on the ground, showing that they had indeed beaten people. The Great Hall was packed with soldiers, nothing but soldiers."

A foreign eyewitness came on the scene shortly after the soldiers emerged. "The street was mobbed. In the middle were about eight hundred soldiers. They had come out of the Great Hall of the People in lines, wearing riot gear, including helmets. Some had cattle prods, some had gas masks hanging out of the backs of their pockets, some had tear gas canisters clipped to their belts. A lot of them had whips, some with spikes at the end shaped like stars, also billy clubs. I saw no guns."

The soldiers were unable either to press forward or retreat into the Great Hall. Two hours later, another eyewitness found the soldiers sitting on the pavement ringed by citizens shouting slogans. "A space had been cleared between the steps of the Great Hall and the first line of soldiers. Red Cross representatives and a student were trying to keep the situation from getting out of hand. An officer was talking to student leaders. The crowd sang *The Internationale* and tried to get the soldiers to sing. The crowd also sang one of the songs of the PLA. It was most extraordinary to see these armed soldiers sitting there sweating in the heat, unable to do anything." The mood deteriorated at the north end, where some soldiers had hit people with their belts and drawn blood.

"Dogs! Creeps!" some women yelled at them. "Who do you think you are? You should die!"

Toward dinnertime, the soldiers were finally allowed to beat an igno-minious retreat. Because the main entrance into the Great Hall was com-pletely blocked by crowds, they had to climb over the fence to get back in.

"Look at those dogs," scoffed the women. "They're jumping over the wall." The soldiers withdrew in humiliation, and the Army's last attempt to rescue distressed units was over.

The events of the night of Friday, June 2, were easily misconstrued as a thwarted attempt to attack the students by stealth. At the time people assumed that these soldiers and weapons were part of a plan to clear the Square before dawn on Saturday. Lightly equipped troops were to run in first under cover of night, skip around the barricades that had stymied the Army's trucks on May 19, and get to the Square. They would form a cordon, and buses coming by other routes would distribute guns to them. While these soldiers prevented more people from joining those already on the Square, better-equipped troops would emerge from the Great Hall to remove the demonstrators inside the cordon. Finally, the soldiers with cables looped around their bodies would use them to pull down the God-dess of Democracy. By morning, it would all be over, and were it not for the vigilance of the citizenry, the plan would have worked. That, at least, is how some chose to make sense of the night's confusing events.

The failure of these troops to carry out such a plan then led others to think that the operation was a Machiavellian plot designed not to succeed. The point was not to take the Square but to start a fight. It was, as one foreign observer assumed, "a deliberate provocation, intended to start a riot and provide an excuse to send armed troops to the rescue." Another foreign resident insisted to me in the same vein that it was inconceivable the PLA would move weapons without adequate protection, unless it intended to lose them. "It makes no sense to send in a truck with arms and let people steal them." He came to the same conclusion: "It was clearly designed so that they could say that people stole weapons, and hence justify the crackdown."

Others argued that what the government wanted was a photo oppor-tunity to show to the world that the Democracy Movement had degener-ated into street fighting. This was Wuerkaixi's view when I interviewed him half a year later. "The Army photographers were out photographing the soldiers that night. The soldiers were not allowed to fight back. When I went back to Beishida, I saw the report on television showing that a lot of soldiers had been hit in the face and so on. The footage showed people among the students, who they were I don't know, stirring up conflict between the soldiers and students. Most likely these were plainclothes police trying to start trouble." In his view, the operation was carefully stage-managed to generate negative publicity against the Movement.

These interpretations of Friday night's events rest on the assumption that the troop movements were intended to discredit the Democracy Movement. In either version, the soldiers were meant to be seen on this moonless night. But the interpretations fall apart if we start from a contrary assumption: that the soldiers were supposed to remain invisible.

These troops and buses were not sent through the streets of Beijing either to capture the Square or provide the government with an excuse to use more violent methods. Their sole purpose was to slip undetected into the Great Hall of the People, and perhaps other locations as well. They had no other orders. They were simply part of the final stage of troop infiltration that the Army had been conducting all week long so as to have as many soldiers as possible in position for the final operation. If an invasion of the Square had been planned for Friday night, there would surely have been more backup PLA activity going on around the city. According to a military source, the main PLA camp west of the city at Mentougou remained quiet from dusk to dawn.

Despite the superficial similarity to what had happened two weeks previously when the people resisted the Army's invasion of the city, the events of Friday night do not add up to an assault. It was never intended that these few troops occupy Tiananmen Square. What the people stumbled upon, and tripped up, were the closing preparations for a much larger, and more deadly, assault that was to come just as the new moon turned.

5

Spilling Blood
(June 3–4)

As Saturday afternoon wound down, quiet descended over Tiananmen Square. It would not have been immediately obvious why the usual din of voices and announcements was easing away. It might have taken a moment to realize that, one by one, the government's powerful public-address speakers, which had blared out anti-Movement messages from the tops of light standards around the Square for weeks, were going mute. About 5:00 P.M., nimble young men started shinnying up the light poles and cutting the wires. People watched from below and applauded every time a line was cut and a speaker went dead. When the last wire was severed, the voice of the government was gone. The students were now on their own. The quiet was like the hush before the concert begins, the stillness preceding the storm.

The government had other loudspeakers atop the state buildings around the Square that the young men couldn't climb to. At 6:00 P.M., the loudspeakers on the Great Hall of the People crackled into life and began to broadcast an emergency announcement. This statement the government would broadcast over and over again throughout the city during the evening. It told the people that they should vacate the streets. Anyone found in the streets was there at his own risk. The government took no responsibility for whatever might happen to those who did not go home. It was a thinly veiled threat of violence, a warning that severe measures were about to be adopted.

Under other circumstances, people might have heeded the warning. But they had been hearing similar statements since April 26, including the impotent announcement of martial law on May 20, and nothing dire had happened. The students and their supporters assumed that whatever new methods the government might be planning to use would be similarly ineffectual. They did not know defeat. How could the Army think of opening fire against its own people, they thought? How could the government condone so severe a measure? Massacre was imaginable but unthinkable. There would have to be some other outcome, the students assumed. A man who refused to identify himself phoned in to the students' headquarters about 4:00 P.M. with a warning that a military operation against the Square was about to begin. The students chose to ignore the warning. There had already been so many.

The unimaginable was being not only thought but put into action. By the time the emergency announcement was echoing over the Square, the Army's offensive against Beijing was already well under way. There would be little attempt at surprise this time. The main factor would be the intimidation of simple weight: well over fifty thousand troops plus armored vehicles would be launched in successive waves from all sides of the city at once, in addition to those already posted inside the capital. Changan Boulevard would again serve as the main access route for troops heading to the Square. Qianmen Street, which leads to the Square from the south, would serve as a secondary access route. Tertiary thrusts would be made along other roads coming in from Fengtai in the southwest, the Fragrant Hills in the northwest, and the International Airport in the northeast. The plan of attack also involved considerable use of the ring roads to move troops around all the compass points. The key locations for securing the city became the intersections along the 2nd Ring Road, particularly the eastern and western overpasses at Jianguomen and Fuxingmen. Because the 2nd Ring Road traces the foundation of the old city wall and its intersections mark the spots where the old gates once stood, the Army's attack took on the formal shape of an ancient military campaign against the capital. Only by securing the key intersections, the "gates," could the invaders take the city.

The commanders of the various group armies brought to the capital met at the headquarters of the Beijing Military Region at 4:00 P.M. to receive final orders. The occupation of Beijing would proceed in three waves of motorized infantry. The first would begin to penetrate the 4th and 3rd ring roads between roughly 5:00 to 6:30, and the second between 7:00 and 8:00. Both waves consisted of large convoys of troop trucks. The soldiers they were transporting would be confined to the trucks and would not carry weapons openly, although their officers would carry pistols. Soldiers in the third wave, launched between roughly 9:00 and 10:30, would carry rifles, and some of these convoys would be accompanied by armed vanguards of PAP riot police. These three waves of troops would be followed by an armored wave about 10:30, and further armored waves after midnight. These moves were to unfold according to a master plan, and none appeared to be contingent on what was happening elsewhere, although the decision to include at least some of the PAP riot squads came as late as 5:30 P.M. Each of the waves seems to have gone ahead as planned regardless of whether the one in front of it failed. The heads of some waves would thus find themselves crashing on the backs of previous waves that had been halted in their path.

Units camped well outside the city's perimeter started to move toward the city about 3:00 P.M. so as to be in position just beyond the 3rd Ring Road when the order to begin the offensive was issued. That order started coming down to some units as early as 5:00 P.M. By this time, however, many of the first columns trying to get into position near the 3rd Ring Road had already run into trouble. The first units to encounter opposition were those coming down from Shahe Airfield north of the city. Several thousand people on their way home from work broke up a truck column from Shahe at the railway crossing at the northeast corner of the 4th Ring Road about 5:00. The column's communications truck was able to withdraw into a senior citizens' home, but then was blockaded there by a crowd of four hundred. A column of fifty-five trucks took up a position just to the north about the same time, again to be surrounded by people getting off work and be told that they should not try to advance because all the intersections were blocked. At 5:50, the commander ordered half his unit to set off, but they were blocked almost the moment they started. Nine trucks at the back end managed to extricate themselves from the blockade and get down to the 4th Ring Road by taking country lanes, only to be blocked by an elderly lady who lay down in the middle of the road. She agreed to get up on condition that they take her with them as a witness to guarantee they would not shoot any students. The 4th Ring Road was fairly open, but crowds increased as the trucks tried to go in further. By 7:30 they had covered only one kilometer.

Another regiment traveling in from Shahe by a different route took a wrong turn onto Airport Road. The bulk of the column got stopped outside the 3rd Ring Road when citizens threw up a hasty barricade of two public buses. Within ten minutes, ten thousand workers coming off shift had them surrounded. Bungling compounded error when it was discovered that the radio in the commander's jeep in the forward command unit didn't work. The head of this column managed to get across the 3rd Ring Road. Its assignment was to take control of the Dongzhimen (East Metropolitan Gate) overpass at the northeast corner of the 2nd Ring Road, but it fell just short of its goal. Shortly after 5:30, citizens overturned a vehicle in the narrow street leading from Airport Road to Dongzhimen and blocked the convoy's further advance.

Other units came from the northeast in the wake of these columns and likewise found their way blocked. A column of forty-one trucks that left

Shahe at 4:58 ended up getting blocked on Airport Road, but further out, at the 4th Ring Road near the Lido Hotel. The colonel in charge of the regiment agreed with student leaders that he should pull his unit back so as not to block foreign guests' access to the airport, but as the students moved the blockade to make way for his unit's withdrawal, the colonel took advantage of the opening to charge forward. In Chinese military lore, this trick of turning a stronger enemy's concession against him is called "using your opponent's spear to attack his shield." The head of this column got no further than the underpass under the 3rd Ring Road when someone dropped a lump of coal from above and smashed the windshield of the commander's jeep. A foreigner came upon the rear section of this regiment about 7:00 P.M. "Eighteen trucks full of soldiers had been immobilized by crowds of people. Someone had thrown broken glass behind them so they couldn't back up, and thousands of people, including housewives with children and old people, jammed in front of them so they couldn't go forward." The one ominous sign was that the soldiers, although apparently unarmed, were wearing combat boots rather than the usual sneakers. Those who had surrounded the soldiers kept up a barrage of pleas and accusations.

"Why are you doing this?" they asked the soldiers.

"If I were in the Army," a young man announced loudly to them, "I'd never fire on the people."

Further in at the 3rd Ring Road overpass, someone in the crowd was telling the colonel in the damaged jeep with the shattered windshield that he should not defend a corrupt government but, instead, should act as had Marshal Zhang Xueliang. The marshal in 1936 had turned against Chiang Kaishek and forced him to cooperate with the Communists in the struggle against Japan, thereby shifting the course of modern history in China.

Shortly after 8:00 P.M., the troops remaining at Shahe left the airfield, turning it over to its regular Air Force guards. An observer estimated that three thousand soldiers set out at this time, though in as much as Shahe had contained about fifteen thousand soldiers the week before, the number may have been higher. Part of the third wave, this force met the same fate coming in along Airport Road. By the time a foreign scholar was driving in from the airport about 11:00 P.M., he found the road choked with troop trucks. "We saw upwards of a hundred open-backed troop trucks, maybe more, plus a few small vans, and nothing was moving. In some places they were clumped along the road, in other spots stretched farther apart, maybe a hundred feet apart, sometimes closer. They were chock full of soldiers, who looked hot and distressed. Students had attached themselves to the trucks, anywhere from three to ten per truck, standing on the running boards and talking to the soldiers in the cabs." At one point a political commissar in one of the trucks out beyond the 4th Ring Road tried to take the initiative and win the crowd over to the Army's side by appealing to the higher authority of the Party.

"Citizens of Beijing," he called out at the top of his voice. "We have come to protect order in the capital, not to suppress the masses. We are obliged to stand with the Party and hope that you understand our position. Don't let bad people in your midst use you to hamper our movements and bring harm to those who are your kith and kin." Given that the commissar stood at the head of thousands of troops poised to enter the city, the plea of innocence won no sympathy.

Shortly after the convoys trying to enter Beijing from the northeast ran into obstacles, units approaching from due east ran into exactly the same reception. The first column stopped on the east side of the city was a small convoy of a dozen or so trucks followed by a officer's jeep coming in along Guangqu Road. The convoy reached the 2nd Ring Road unhindered at Guangqumen, then went north to the Jianguomen overpass, where it tried to get up onto Changan Boulevard. Its goal may have been to secure the overpass. About 5:50, directly beneath the balconies of the diplomatic compound at Jianguomenwai, thirty to forty resolute individuals managed to stop this column on the exit ramp and let the air out of the tires of the lead vehicles. Two or three trucks made it onto the overpass, but the rest were stuck on the ramp or in the exit lane underneath the bridge. Within minutes the trucks were surrounded by several thousand people. They sat on the hoods of the trucks or clambered onto the tailgates to talk to the soldiers. Within half an hour, the crowd trebled in size. The soldiers could do nothing but listen in silence to the endless explanations people gave regarding the just demands of the students. The occupants of one truck got out and sat on the ground, but the rest remained standing in their trucks.

A foreign businessman who speaks fluent Chinese walked up onto the overpass ten minutes later. Some people quickly approached him to speak on their behalf, as though his status as a foreigner might give him some authority with the officer in charge.

"You must talk to the commander," they said, "because he won't listen to us." The commander was sitting in the jeep at the rear of the convoy, a quiet, professional-looking soldier of about forty. They talked for about twenty minutes.

"You are a foreigner," the officer pointed out, "and foreigners have been warned not to be involved in anything."

"Well, let me tell you that I live right at this intersection. I've lived here for two years, and I've lived in China altogether for four years. You can call me a foreigner, but I'm also a resident of Beijing. I work in foreign trade. I'd like you to know that, whatever else these people are telling you, having soldiers coming into the city is going to have a very bad effect on trade. I have come to talk to you because I am very concerned about what you are doing here. When you bring soldiers into a city in large numbers with weapons, somebody is bound to get hurt. It's not good for the people, and it's not good for trade."

The commander was willing to listen but otherwise said little. Recalling this conversation two months later, the businessman observed: "Basically, he didn't know what to do. He didn't say this, but there he was with his trucks so surrounded by people that they couldn't move. People climbed the sides of the trucks and stood up with their faces right next to the soldiers and talked to them. There was a continuous conversation going on with all the soldiers who were at the edges of their trucks. Some people were very emotional, almost rude; others were polite. But they were all asking the soldiers the same thing: get out."

At 6:15, another column of thirty-five troop trucks came up to the Jianguomen overpass from the same direction. Part of the crowd detached itself and went down onto the ring road to intercept the trucks. The first ten vehicles in the column were able to slip past Jianguomen and continue north, only to be stopped by another crowd at the next overpass, Chaoyangmen (Sun-facing Gate). Like the first convoy, the new arrivals at Jianguomen were obliged to listen to the lectures and scoldings of the crowd. One or both of these convoys were said to be from the Shenyang Military Region, probably the 16th Group Army, possibly the 39th. Four hours later, bystanders told a foreign journalist that the soldiers appeared to have been told nothing about what was going on in Beijing.

The group army that was to mount the second wave coming in from the east set off from Tongxian at 5:50 P.M. in four separate columns traveling by different routes. (By 7:00, crowds in Tongxian had rendered many of the roads leading west to Beijing impassable.) One part of this force came in along the main highway from Tongxian that feeds directly into Changan Boulevard. The Jianguomen overpass lay directly in its path, but even before it got there, the column was broken up by crowds further east. Some twenty to thirty trucks were stopped at Dabeiyao, where Changan crosses the 3rd Ring Road. Thirty-two made it through, but short of running down the fifteen thousand people now posted on the overpass, it had no chance. As the lead trucks reached Jianguomen at 7:10, their way was blocked. The major in charge of the unit retreated to one of the skyscrapers facing Changan and set up a temporary command post there.

Another unit ordered into action at 6:00 P.M. as part of the second wave was likewise broken up on its approach to Jianguomen from the southeast. The head of the column made it to the overpass only to be stopped dead, while the the rear sections were stopped along the 3rd Ring Road at the two main intersections south of Dabeiyao.

The third wave of troop trucks coming in from the east were even less successful in penetrating the citizens' defenses. A major convoy of possibly as many as 150 trucks set out from Sanjianfang Airfield northwest of Tongxian before 9:00 P.M. but was stopped by a four-vehicle barricade workers had pulled across the Tongxian-Beijing Highway halfway to Beijing. The column stretched from this intersection all the way back to the bridge just west of Tongxian. The convoy remined immobilized for seven hours until a fleet of four APCs plowed a path through the barricade at 4:00 A.M.

In all these locations, citizens used techniques they had learned and perfected since May 19. Break the column's momentum with bodies or buses, force the lead vehicle to come to a halt, and then incapacitate it at its four Achilles' heels, the air valves on the tires. Box in the rest of the column so that the other trucks can't get around the lead vehicle, then deflate their tires as well. Finally, open the hoods and disconnect wires or rip out engine parts. It went like clockwork. Saturday night began to look like a repeat of every other attempt the Army had made thus far to drive into Beijing. By the time the third wave was launched, at least two million people were in the streets, and everywhere the Army was stopped in its tracks.

The only attempt to relieve the soldiers at Jianguomen occurred about 6:45, when four trucks carrying men in blue factory jackets and yellow hard hats came down the 2nd Ring Road. The four were followed shortly thereafter by another three. As the trucks approached, the crowd on the overpass cheered. They thought reinforcements were coming from an industrial enterprise. But something about the convoy looked wrong and the crowd became suspicious. An eyewitness from the diplomatic compound nearby watched people pour down onto the ring road and prevent the trucks from progressing. "One truck got away—they were too late to catch it—but the other three they caught and let the air out of their tires. Suddenly, all the hard hats flew into the air. Then there was a great smashing sound: pop bottles being carried in the trucks were thrown out and smashed on the road. Some scuffling started when the second set of three trucks came up a little bit later. Those trucks too were stopped, though I think one or two were able to reverse quickly and get away."

Another eyewitness arrived on the scene about 7:00. "We got there just as they started to jump off the backs of the trucks, but they were surrounded by people. They had long sticks, but they never tried to fight their way into the crowd. They realized the numbers were just too large to start anything. By this time the truck tires were flat and people had opened the hoods and pulled pieces out of the engines. They were quickly dispatched by the civilians. They took off up the ring road, chased by a crowd."

Despite appearances, these were not workers but members of a special police unit used in undercover operations. The unit was known and detested; it had played the same role in suppressing the demonstration at Tiananmen on April 5, 1976. On that occasion its members had posed as workers from Capital Iron and Steel. Their assignment at Jianguomen appears to have been to open a path for the beleaguered Army trucks in the fraudulent name of the working class. They were smelled out too quickly to achieve their purpose. Members of the crowd clambered aboard, threw their staves to the crowd to destroy, and chased them off. An elderly man standing on the overpass told a foreigner that the people had nothing against ordinary soldiers and policemen, but they truly hated the special unit. "Anyway, we have the soldiers surrounded. They can't even piss without our permission."

The scene of staves being thrown from the trucks to the crowd reappeared in July in one of the propaganda videos the government compiled to sell its version of events. The voice-over says the time was 5:00 P.M., the place Tiananmen Square, and the occasion the distribution of weapons to the crowd by members of the Workers' Autonomous Union. None of these statements is true. The scene at Jianguomen had been transposed. The footage was so tightly framed that it did not give away background, and one couldn't really tell where it was—but the light was wrong, too pale for 5:00 P.M. This editorial sleight of hand was a second try at deception. Perhaps, thought the film editors, if the people at the time weren't fooled by appearances, maybe the television audience could be duped into believing that the Workers' Autonomous Union was arming workers in the Square. It would be visual proof of the government's assertion that the people initiated the violence. This forged incident of workers distributing weapons on the Square at 5:00 P.M. has now entered the government's official account of the suppression. It is one of the few outright lies in the government's story.

As the evening wore on, East Changan Boulevard became impassable to ordinary traffic. At the 3rd Ring Road, the Dabeiyao intersection was now blocked by a ten-bus roadblock three buses deep, leaving only a narrow bicycle lane through the center. The barricade on the crowded Jianguomen overpass was flimsier. Only two buses were blocking the western exit of the overpass, but other obstacles were being thrown up between there and the Square. A foreign journalist coming east along Changan about 10:00 P.M. encountered "barricades in the street the whole way down. I also remember seeing two or three old women running with big sharpened sticks, five or six feet long, handing them out to students. Generally people were moving eastward toward Jianguomen. We also saw a couple of dozen young workers—we were told they were workers, not students—all dressed in the same T-shirts. They were the students' protectors," a Dare-to-Die Squad. Half an hour later, the same journalist was on the Jianguomen overpass and commented to a bystander that it was late.

"Yes, normally I'd be in bed now," the woman said, "but I have to stay here, and I'll stay all night if I have to, to make sure the soldiers don't go toward Tiananmen Square."

By this time, the people at Jianguomen knew that these soldiers had AK-47s. They weren't strapped to their backs or otherwise visible, but they could be seen under tarpaulins at the backs of some of the trucks. When one foreigner who was not aware of the guns challenged the soldiers over the belts some had coiled in their hands, a witty bystander quickly retorted, "What do you think they've got guns for?" As yet they had no order to use them.

First- and second-wave convoys that broke on the east and northeast sides of the city were also launched from the north and south, with even less success. The first convoy from the north came down College Road, and then crossed east along the 3rd Ring Road to the next intersection at Beitaipingzhuang, where it intended to turn south and head into the city. An eyewitness was in a restaurant at the market at Beitaipingzhuang when the convoy approached shortly after 6:00.

"We had just finished dinner and I was sitting by a window when I saw people flooding into the streets through the market stalls. I got up and looked out the window and shouted, 'The Army's coming!' Everybody ran out of the restaurant. By the time we got out on the streets, thousands had surrounded the convoy. There were a couple of buses, followed by about ten open-topped troop trucks. Soldiers were standing in the trucks, so we saw them from shoulder level up and I don't recall seeing weapons." What surprised this eyewitness was that the commander didn't seem flustered about being stopped. "People argued with him and talked to the soldiers, explaining that the Democracy Movement was patriotic. They offered the soldiers ice cream and drinks, though for a long time the soldiers refused to accept anything. Some of the soldiers started to cry as the crowd pressed food and water on them." After about half an hour, the confrontation ended peacefully. "As the trucks turned around, everybody cheered. The people did not clear the streets, however, because they didn't trust the Army not to double back on them. They moved as a crowd with the buses and trucks around the intersection and escorted them west about 500 meters to the next intersection, from where they went north out of the city."

A second convoy, dispatched from Shahe Airfield at 8:00 P.M., made its way down College Road later in the evening, but did not penetrate the 3rd Ring Road. Another contingent coming from Shahe was stopped well north of the city. The back end extricated itself from the blockade and headed off over small country roads to avoid detection. Challenged at every turn by the peasants, these soldiers ended up taking off their uniforms, removing the military plates from their vehicles, and driving without lights one truck at a time in order to evade detection.

Units also came down on the northwest quarter of the city from the military base in the Fragrant Hills out beyond the Summer Palace. A convoy of eighteen to twenty trucks was spotted at 7:45 P.M. moving southward from that area. Part of a different column of about seven hundred soldiers coming from the northwest made it almost as far as People's University before being stopped. Three or four trucks of PLA equipment, one carrying APC treads, were abandoned and left to be burned. (Neither People's not any other university was targeted for attack that night. Located in the northwestern part of the city, Beijing's leading centers of higher education were vulnerable to attack from the northwest, yet the PLA showed no interest in approaching them.)

The advance from the south met as much resistance as the convoys coming from the north, but it proceeded more rapidly because of larger and more direct roads covering the twenty kilometers between Nanyuan Airfield and the south end of the city. The first units received their orders at 4:15 P.M. to move north; the next set of orders was issued at 5:00. Official accounts describe the experiences of some of these units. One company of 122 men in this first wave from the south set out at 5:00 on secondary roads west of the main highway and managed to cross the 3rd Ring Road before getting blocked. They were forced back to the ring road at Majiabao and found themselves surrounded by thousands of people who held them there until midnight. A regiment of 880 men that set out from Nanyuan at 5:20 via roads to the east of the highway encountered even greater opposition. The official account says they were blocked sixteen times between the airfield and the Square.

The second wave involved larger and more determined detachments that the first. A large column left Nanyuan at 8:00 P.M. as part of the second wave. The elite 15th Airborne formed the core of this column, with head and tail consisting of units from other group armies. This column took the more direct route straight toward Yongdingmen (Gate of Eternal Stability), where the south highway crosses the old moat to the southern city. These soldiers met their first resistance at 8:25, five kilometers south of the 3rd Ring Road. Part of the convoy maintained good security around its trucks and was able to push ahead gradually and cross the ring road at the Muxiyuan traffic circle, but it was eventually hobbled by barricades thrown across its path farther ahead. The rest of the convoy fell short of the traffic circle. These soldiers failed to keep the crowds away from the trucks and their tires were slashed.

Another regiment coming from a more distant camp south of the city also found its way blocked at Muxiyuan and shifted west, trying to cross the moat at the next bridge west of Yongdingmen and go up Taiping Street. At 10:30, three foreign tourists came out of their hotel nearby and found part of this regiment stopped on the bridge. "A convoy of about ten military trucks had tried to come north across the bridge at the bottom of Taiping Street and was stopped. Half the column was on the bridge; the other half hadn't made it over. There must have been two or three hundred soldiers. They wore gear that looked like they were ready for combat, including helmets, and were armed at that point. The barricade was made up of cars and cement blocks, not very substantial."

The three foreigners turned and went north by pedicab, heading in the direction of the Square. In less than a kilometer they found a second body of troops. "Taiping Street was completely blocked by big trucks. The column was off to one side, and they couldn't have got through. People were in heated debate and the soldiers were not smiling. There were not as many trucks in this column, but there was a communications vehicle there, a jeep with a lot of antennas. It seemed that they were both part of the same column as those we had already seen. This was the front end, where decisions were being made."

As the foreigners went around these soldiers and made their way north to the Square, other second-wave troops coming in from the southwest were making greater headway. A contingent of twelve hundred soldiers plus two trucks coming from the outer suburb of Fengtai was only a few blocks from the Square when it got stopped on West Qianmen Street. This small group quickly expanded as hundreds of students and workers raced out of the Square to lend support. An articulated bus was added to the barricade across West Qianmen Street. The soldiers eventually withdrew, to the cheers of the crowd.

Prior to the third wave, then, the Army had advanced on the city in coordinated moves from all quarters, but had been frustrated at every location by crowds of people adamant that soldiers not enter their city. In most spots, the barricades were flimsy structures thrown up at short notice. Without the order to go ahead at any cost, which had still not come down, the troop convoys could not get past them. The majority of first- and second-wave troops did not even make it past the 3rd Ring Road.

Given this resistance, the strategists were right to have infiltrated soldiers into the city before the attack began. It began to bring these reserves into play at roughly the time the third wave was starting. What the Army was trying to achieve by the few maneuvers that were detected, however, is a puzzle. For instance, a contingent of troops in fatigues and helmets, reportedly from the Palace Guard, erupted from the Great Hall of the People precisely at 9:00 P.M. They were immediately surrounded. Only an hour later were they allowed to withdraw back into the Great Hall, with the commander promising they would not enter the Square for the next forty-eight hours. They accomplished nothing, and one has to assume their commander broke his word.

Just as the Palace Guard went back into the Great Hall, another unit suddenly appeared in close proximity to the Square. A column of soldiers jogged by the Beijing Hotel, apparently on their way to the east gate of the Forbidden City. Twenty-four of them, stripped of caps and jackets, were apprehended by bystanders after striking out with belts at foreign cameramen. They were escorted away by student monitors. Anyone standing in the front parking lot of the Beijing Hotel at 10:00 P.M. assumed that the fiasco of the previous night was being reenacted, as unarmed citizens stopped, humiliated, and forced back unarmed soldiers.

If there was an ominous sign that Friday night's scenario would not be repeated, it was the presence of plainclothes agents in and about the Square Saturday night. Most were detected only by foreigners, because one of their tasks was to get foreigners away from the impending military action. Undercover agents had been working in the crowds during the late afternoon. Some stirred up street fights outside Zhongnanhai at 4:40 P.M. Others had been on the Square gathering intelligence (at least one fell into citizens' hands and was escorted to the students' headquarters on the Monument). When the loudspeakers controlled by the Workers' Autonomous Union at the northwest corner of the Square at 4:30 broke into a hysterical call to "fight counter-revolutionary violence with revolutionary violence," it struck many that government agents had taken over the union in the hope of goading workers into violence and thereby subverting the Movement. As evening came, plainclothes officers began to warn foreigners to get off the streets. The Chinese government did not want foreigners caught in crossfire, nor did it want them to observe actions that were supposed to be invisible. A foreign teacher on the Square was approached by a young man about 10:00 P.M.

"You should leave now, because it is too dangerous," he said in excellent English. "You should go to a hotel. The Army is going to come in."

The teacher's first reaction was to shrug this off as yet another rumor in a month of false predictions. "It wasn't immediately apparent to me that he was out of place. He looked like a student. But later I wondered why there happened to be somebody who spoke good English in the crowd warning foreigners to go away. It struck me that he was on assignment." Other foreigners have reported similar experiences on the streets around Tiananmen Square on Saturday evening. Few had the foresight to realize that these men were preparing the ground for attack.

A Lebanese student found himself in an ambiguous situation toward midnight when he was caught in a crowd on a street south of the Beijing Zoo. He was taking photographs, in clear contravention of martial law regulations, when the people near him started to run. Fearing police action, he put the camera away and decided to head home.

"Grab him!" someone yelled.

"I felt so scared," he recollected later. "Without thinking I started to run, but after five steps in my sandals I found I couldn't run fast, so I took them off and carried them in one hand and the camera in the other. Ten meters ahead of me on the sidewalk I saw this guy put his leg out in front of me. I did not want to stop—I was scared of what was behind me—and thought I could just push past him, but my hands weren't free and I fell. At least twenty people clapped their hands on me. I stood up but couldn't even move."

"Who are you? Why are you here?" they screamed.

"Listen to me! Listen! Listen! I'm a foreign student."

"He's a liar! He's from Xinjiang!" They suspected that he was a Uighur from Chinese Central Asia working undercover for the government. Fortunately, one of the people tried speaking English to him, and he could prove he was indeed a foreigner. The crowd cooled down.

"Show us your student card."

"Look, I'm a student," he said, holding it up. "I support you!" Suddenly the mood shifted and everyone started applauding.

"Why did you run away?" one of them asked.

"Somebody yelled 'Grab him!' and I ran. I was afraid he was with the Army. Had I known he was a citizen, I wouldn't have run away."

"When you run away, people get suspicious and they will go after you." Chastened by this lesson in crowd psychology, the foreign student went home, still unsure who had started the pursuit. The next day, after the Massacre, a neighbor dropped by his apartment.

"It wasn't the people who tried to grab you last night," she confided. "It was the police. They couldn't say that they were with the Army, or the people would have killed them. The people wouldn't have harmed you. I have a son in the Communist Party, but even so we hate them. Before it was alright, the Communist Party were our own people. Even if our lives were hard, we had nothing against them. But now that they have turned their arms against us, we hate them."

It was on the west side of the city that the arms of the Party were first turned on the people. The main force used to clear the Square Sunday morning came from this direction. Units based west of the city appear not to have been part of the first wave. Only at 5:00 P.M. did the first units receive their marching orders. Two infantry divisions, followed by an armored division, an artillery brigade, and an engineer regiment, were ordered to set out at 8:00 P.M. by different routes and establish a forward base at the Military Museum, a kilometer inside the 3rd Ring Road, before 10:00. Other units, some of them stationed as far west as the outlying town of Gucheng, would then follow in their path as the main third-wave force. Anticipating opposition, the lead units were to be protected by a PAP vanguard of six hundred fully armed riot police. The PAP presence was a sign that the western axis was key. No column coming in from any other direction was given this support.

The column met its first serious opposition at the east side of the Gongzhufen traffic circle on the 3rd Ring Road. The troops tried to press forward through the protesters standing in their way. The mood turned ugly. "Get the fuck out of here, soldiers," some young men yelled. They started to pitch rocks and bricks, shattering a few windshields. Progress was painfully slow, and the troops were able to push ahead only by returning the hail of rocks. By 9:00, the head of the column was approaching the Military Museum, but the other units still remained beyond Gongzhufen, outside the 3rd Ring Road. A contingent of police tried to clear a path for the several dozen troop trucks stretched along the two long blocks west of Gongzhufen, without success. When the commander of the Beijing Military Region reached the tail end of that truck convoy at 9:55, everything was at a standstill. It would take his jeep another hour to cover the next four kilometers between there and the Military Museum.

At 9:20 P.M., the force at the Military Museum began to push its way further east. Thousands of workers and students formed a slowly retreating human wall fifteen rows deep across Changan Boulevard. Opposite them, across a hundred-meter no-man's-land littered with stones and broken bottles, stood at least five thousand soldiers, helmeted but unarmed. "We will put an end to the turmoil," they chanted. "Soldiers don't stop till the turmoil stops!" A few wielded billy clubs and leather belts, and the rest threw the stones that demonstrators had thrown at them or stacked along the side of the road. The PAP escorts lining each side of the column's flanks carried AK-47s.

The first warning shots were fired into the air about 10:00 P.M., loud bursts of gunfire meant to intimidate the demonstrators and keep them well back from the troops. The demonstrators flinched but did not flee. "Put down your arms!" they chanted.

An officer in a jeep picked up his megaphone. "Charge, you bunch of cowards!" he ordered the soldiers in front of him. "Sweep away this trash!"

The soldiers followed orders by throwing several dozen stun grenades. These exploded as they hit the ground in front of the demonstrators, making a tremendous crash but not injuring anyone. It was 10:35. The people in the crowd turned and fled, tripping over abandoned bicycles and scaling walls to get out of the advancing column's way. Within minutes, young men were back in the street again, throwing stones at the soldiers and stopping them after they had advanced barely a hundred meters.

"Fascists!" the people screamed. "Stop!"

By slow steps, the column continued to push its way east toward the bridge at Muxidi. A Chinese eyewitness arrived at Muxidi ten minutes later, drawn from his home by the sound of explosions. For half an hour he saw no soldiers. The Army had not yet made it to the bridge. "All of sudden someone yelled, 'Make way, make way!' A row of three-wheeled flatbed bicycles were taking wounded people out. I saw two guys who were definitely shot dead, and I saw two or three others with gunshot wounds." One of them was a student from Beijing University who had stepped forward into no-man's-land to negotiate with the soldiers. He died of a head wound shortly after reaching Fuxing Hospital nearby. As of 11:00 P.M., the killing had started. The soldiers with the AK-47s were not just firing into the air. Now they were taking aim.

The order to fire was requested by a junior officer, but it was not given by him, despite rumors to the contrary. Field officers do not have that authority; nor do unit commanders. It had to come from the Central Military Commission, almost certainly from its chairman, Deng Xiaoping. Vice-chairman Yang Shangkun may have issued the formal order to shoot, as the Beijing rumor mill insists—he certainly appears to have been the man who approved the plans for the June 3 assault and monitored its development through the night from Zhongnanhai. But the order to open fire had to originate with his superior on the CMC, Deng Xiaoping.

Not all lower-level commanders who by this time were radioing from many locations around the city for permission to use live ammunition received authorization. An official account of one regiment's attempt to enter Beijing from the southwest states that a soldier asked his regimental commanding officer and political commissar for permission to open fire at 10:48 in face of overwhelming popular resistance. The account does not say whether this request was passed to higher authorities, but permission was not granted. As the night wore on, though, more and more units were cleared to "go ahead at any cost."

Although PAP were now aiming their AK-47s into the crowds at Muxidi, casualties from direct rifle fire at this early stage were few. The Army continued to use nonlethal tactics whenever possible, including more stun grenades. Finally, the head of the soldiers' column reached the bridge at Muxidi where Changan Boulevard crosses the canal flowing out of Diaoyutai, the residence for visiting heads of state. There they met a formidable barricade. Four trolley buses, two dump trucks, and several minibuses had been dragged across the road. Young men standing on top of these vehicles hurled rocks and bottles handed up to them by older men and women. A young worker in overalls stuffed a burning rag into the fuel tank of one of the trucks and slowly the wall of vehicles turned into a wall of fire. The Army was stopped dead in its tracks. Journalist Pierre Hurel was at the scene:

> In front of the flaming barricade, alone, facing the soldiers, four students, their feet planted wide, crack the heavy air with the sound of their waving scarlet banners. In an unbelievable gesture of defiance, they are naked, martyrs before a sea of soldiers in brown combat helmets tense with anger. The silk of their university banners gleams in the fire's light, and behind them, the crowd, waiting for the worst, applauds them. It is 11:30. And for the first time tonight, the soldiers have had to pull back.

Some soldiers, injured by flying rocks, are taken to hospital.

The soldiers' retreat proves to be only a brief pause for the military juggernaut rolling eastward toward the Square. A Chinese eyewitness watches the soldiers resume their push onto the Muxidi bridge from his vantage point on the smaller bicycle bridge a short distance to the north. "By this time the trolley bus on the main bridge is on fire, and in the light of the flames I see about two hundred Armed Police with helmets but no guns, just sticks, trying to break through the line. People on the main street are throwing rocks. The people on the bicycle bridge are calling out at the tops of their voices. Finally the Armed Police break through, and trucks pull up behind them. They cross the bridge and get full control of the intersection."

"Bandits!" people scream. "Bandits!"

Now the guns in the hands of the PLA are being fired, and the shooting begins in earnest. From the opposite side of the street, Hurel sees several hundred men armed with AK-47s leap over the vehicles, form up in a line, and position themselves to cover the rest of the column. As the troops behind them throw stones, marksmen on the barricade start firing.

> In front of the burning trucks, the four banners collapse. Behind, a dozen fleeing students are struck by bullets in the back. Right near me at the end of the canal, a young boy, maybe fifteen, picks up one more stone to throw one more time. In a fraction of a second, his white T-shirt is stained with red in the middle of his chest. From where I am hiding behind a tree, I see the marksman, his back up against the side of a bus, his gun at his side, and he starts firing again. I run across open ground toward the shadows where several hundred Chinese are yelling in flight. Then something crashes above my hip.

Hurel discovers he has been hit, but is able to get back to his car and driver close by. "My driver is terrified. Around the car, a dozen Chinese bang on the locked doors. The wounded. I get into the front seat and three young men climb into the back. The fingers of one of them have been torn off. Another, an adolescent, helps his brother whose stomach had been opened by a bullet. When I turn to look at them, the pain cuts me in two. The seat is sticky with my blood and my shirt is stained red." As the car makes its way through the crowds, they pass dozens of injured people. "In front of us, a woman is carrying her unconscious husband on a three-wheeled bicycle. Jostled by the fleeing crowd, she overturns by the curb. The young man in the seat behind her vomits blood." Hurel's young passengers are afraid to show up in a government hospital with bullet wounds. They elect to be taken home. Hurel gets treated at a foreigners' clinic.

Shortly after this, at 12:14 A.M., a BBC correspondent approached the area from the east. She and her colleagues were intent on getting to Fuxing Hospital, the hospital closest to Muxidi that received the first wave of casualties. "There was so much gunfire between us and the hospital I never got there. In front of us was a horrendous scene, one of great confusion . . . , a great deal of shooting, quite a number of corpses, people being shot in front of us. There was panic and a lot of people running around, unaware of the seriousness of the situation." She gave up at 1:00 A.M. and headed instead for another hospital "We came across a group of people with a woman very severely injured. We took her in the car—one of the few vehicles on the road that night—to hospital, and there I got some idea of the scale of things, seeing something like 120 people come past me, severely injured, with some dead."

People elsewhere in the city at first refused to believe that the Army had opened fire or that the soldiers, even if they did fire, were using live ammunition. At the north end of the city at this time, no one knew what was happening at Muxidi. Citizens blocking soldiers at the Madian overpass on the 3rd Ring Road were still speaking to them in a friendly manner at 11:30, offering cigarettes and soft drinks. The soldiers, who spoke in local dialect, seemed both aware of recent political developments and sympathetic to the students.

"We understand the students' demands," they said. "The students only want the best for our country. We will not use force against them."

A British student listening to this exchange left Madian shortly after midnight. To her surprise, a young Chinese man fell into step with her and struck up a conversation. He revealed that he was a recently demobilized soldier who had served as a security guard for the top leadership. Such a post restricted him from talking to any foreigner for at least two years and from going abroad for five, but he wanted to know how she felt about what was taking place. He also seemed keen to set her straight on the real character of the PLA.

"The propaganda about this being a *people's* army is absolute rubbish," he said. "In actual fact, we are supposed to keep the people *under our control*. The people and the PLA are pulling in opposite directions and have no common goal. Most soldiers are uneducated and lack the ability to think for themselves. I suppose that's why they were glad to get rid of me. I can think for myself and started to realize that the soldiers around me were nothing more than cannon fodder."

As he talked, they walked toward Beijing Normal University. She asked him what he thought the outcome would be, adding that she did not expect anything much to happen that night.

"Don't be so sure about that," he replied. "I suggest you get back onto the campus now."

They could hear gunshots in the distance. It was getting on toward 1:00 A.M.

"This is just the start of it. It's going to be bad tonight." He stood up. "I'm going home now, too."

Violence broke upon downtown Beijing not in the form of rifle fire, but in the sullen shape of armored personnel carriers running at full throttle. These APCs were part of a wave of armored units, the first of which were ordered into battle at 10:30. The role of the APCs was both to drive a wedge through the crowds that had blocked the soldiers coming in by truck and to transport soldiers into the Square. Several got separated from their units, however, and ended up charging up and down the streets of Beijing, creating a mayhem of noise and alarm.

APC 332 was the first to reach the Square. It stopped at the southeast corner at 11:32, completely lost and with a radio that wasn't working. The vehicle circled the Square twice, drove off first along East and then West Qianmen Street in search of its unit, came back to the Square, then finally drove around to the far side of the Great Hall of the People to wait. A crowd surrounded the vehicle, but the commander through persuasion was able to keep them from destroying his APC.

At almost the same time, APC 339 raced across the north end of

Tiananmen Square heading east on Changan Boulevard. A Beida student saw the APC "fly across the top of the Square. I assumed it was clearing the way for the Army. It was being driven at a very high speed, paying absolutely no attention to whether anyone was in its path. If you were in front of it, it was just going to run you over. A lot of people were running after it." Then a command-post vehicle, which can be identified as such by its double-zero number APC 003, followed APC 339 by a few minutes. Traveling at a top speed of almost eighty km/h, the two vehicles created an uproar all along Changan. Apparently they were looking for the Square but missed it as they drove by.

APC 339 charged east and plowed across the Jianguomen overpass, where thousands had gathered to block troop trucks. Because the foreign diplomatic compounds were just to the east, many foreigners witnessed the APC's movements. One standing on the south side of the bridge turned when he heard "a strange noise" and saw an APC coming straight at the bridge along Changan Boulevard. "It bypassed the first barricade, which was simply two articulated buses across the road, but with enough space on the south side for it to climb part way up on the sidewalk and get through. It came crashing over the rows of bicycles on the bridge, passed within about ten feet of me, and managed to squeeze its way between the Army vehicles partially blocking the east end of the overpass. People just melted into the sides of the bridge." As another eyewitness put it, "the APC sent people scattering like water from a boat. Fortunately, nobody was run over."

APC 339 continued east almost as far as the 3rd Ring Road, pursued by a brigade of angry cyclists. By then the commander realized that he had overshot his target. He wheeled his vehicle around and raced back in the direction from which he had come. As the APC passed in front of the Qijiayuan diplomatic compound a kilometer east of Jianguomen, a cyclist threw himself down on the road twenty meters in front of its path to force it to stop. At the speed the APC was traveling, the man didn't have a chance. The vehicle struck him full on, killing him instantly, and then continued west without pausing. A foreign policeman who saw the incident from his apartment went down to see if anything could be done, but by the time he got down to the street, an unmarked white van had already pulled up to take the body away. The door at the back closed and the van took off at high speed. Government eyes were watching all this happen. No evidence should be left behind.

The crowd back on the Jianguomen overpass was electrified by the sudden passage of the APC. "The student organizers of the blockade decided to reinforce the barricade. They pushed trucks full of troops around on the bridge to make it almost impossible for any vehicle to get through. Then suddenly the APC was coming back. It slowed down a bit, and people threw bricks at it. One civilian actually got on top of it. It couldn't get through, so it simply smashed into the barricade." Another eyewitness watched APC 339 plow its way across the overpass. "The APC hit one of the five troop trucks in the middle of the overpass. The truck tipped over and soldiers spilled out. The crowd screamed in a frenzy. Lying on the other side of the truck was a civilian's body. He had been crushed, his head was crushed, his blood and brains quite visible on the sidewalk." For yet another eyewitness, this death has become a recurring nightmare. "The civilian's head was popped like a grape. I'd never seen anything like that. It still haunts me."

"The APC had to slow down considerably to get through, and people were hurling the most awful abuse at it, as well as whatever they could get their hands on. Again it clattered over the bicycles and squeezed its way past the bus and headed back toward the Square." It zoomed off, leaving a shower of broken bottles and a hundred crushed bicycles in its wake. The demonstrators hoisted the truck back up on its wheels, and then turned on the soldiers. They surrounded them, howling at them for what the Army had done. Some even tried to pull the clothes off one of the soldiers to shame him and degrade the PLA uniform.

"Look, your own army's killing you!" they screamed. "Why don't you fight back?"

At least three soldiers were wounded when the APC crashed into the troop truck, one seriously. They were carried on stretchers down below the overpass for protection. An ambulance soon came along the 2nd Ring Road to retrieve them but was hindered from picking up the soldiers by a hail of rocks. The crowd was enraged that the ambulance was there to pick up only the soldiers and not the dead civilian. His corpse was left as a bleak memorial of what had happened. It would remain untouched on the overpass for several days. His identity remains unknown. All that could be identified about him was his boots, which came from Shanghai.

Unlike APC 339 ahead of it, APC 003 stopped short of the Jianguomen overpass. Its commander, Colonel Dong Xigang, realized that he had overshot the Square before he got to Jianguomen. He turned down to the Railway Station, circled up to Dongdan, then drove back into the Square, arriving at what Dong calls his "appointed location" at the northeast corner of the Square at 12:20 A.M. By this time his radio operator had lost contact with APC 339. As it entered the Square, the APC had to drive over metal traffic dividers dragged across Changan. The episcopes were damaged and the driver failed to negotiate his way around the obstacles. The left tread caught in one of the traffic dividers. The APC stopped, got going again, but couldn't disengage itself from the railing. After weaving another hundred meters or so down Changan, it came to a stop at the north end of the Square. Instantly, a crowd of young men surrounded it and attacked it with whatever they had in their hands. One man got on top of the vehicle and tried to force the hatch open with a crowbar. Then the crowd tried burning the occupants out. They set fire to pieces of wood underneath the

APC, broke Molotov cocktails over the top, and draped it with a flaming quilt.

What the occupants of APC 003 either did not know or were too scared to trust is that an APC provides adequate protection against this sort of crowd attack. The inner compartment of an APC is pure steel one inch thick. The exterior paint and rubber seals may burn, heating up the walls inside, but the crew is safe so long as the hatch remains closed. And without explosives, it is almost impossible to pry open a properly secured hatch from the outside. The only way to ignite an APC is by smashing a Molotov cocktail on the front intake grill and getting fuel into the carbure-tor, but that did not happen. According to the published testimony of crew members, fires somehow started inside the cabin and the crew pan-icked. Colonel Dong decided to open the hatch and reason with the crowd, but he and all eight crew members were set upon by the crowd. Badly beaten, they were rescued by students and taken to hospital. APC 003 was left to burn.

When APC 339 reached Tiananmen Square shortly thereafter, people again threw rocks and Molotov cocktails. It turned away from the flaming vehicle before it and headed down the east side of the Square, but it too was stopped. Eyewitnesses reported seeing a volley of heavy-caliber tracer shots rocketing over Tiananmen at an angle of about 45 degrees from the southeast. It seems that a panicking crew member fired the machine gun mounted on top of the APC to warn away the crowds. Like the crew members of APC 003, the soldiers in APC 339 abandoned their vehicle and left it to the crowds to torch.

These vehicles were part of a larger armored unit that came into the city from the southeast, crossing the southern moat at the Puhuangyu Bridge southeast of the Temple of Heaven. Their assignment was to take up positions at the north end of the Square directly in front of the Forbid-den City without injuring anyone. But somehow the vehicles split up, got lost, and panicked. Error reigned. Colonel Dong had been seconded from his own vehicle (APC 229) only at 7:00 that evening and was inadequately briefed about the operation. Radio contact with the other vehicles was lost. In addition, the driver of APC 003 has admitted that he had never driven in Beijing before. Given the limited vision that the three small episcopes in an APC provide, especially at night, it is not surprising that the driver lost his bearings and sailed past the Square without realizing where he was. Very possibly he had no map: the officer who took command of Colonel Dong's APC 229 (who in any case wasn't even an APC commander but a medical doctor) has revealed that the head vehicle in his column dispatched at 11:30 "didn't know the route of attack, and also didn't have a map of the Beijing region." Whatever other errors were involved, the unintended consequences of sending APCs 003 and 339 through the streets of Beijing were critical. The thundering machines put the entire city on high alert, killed and injured several civilians and soldiers, and precipitated a rage that goaded many people into blind violence.

APCs were now in motion all around the city, attempting to work their way toward the Square in the company of truck columns. The main force coming in from the west, which toward midnight pressed through another flaming barricade at the Fuxingmen overpass, was joined by a column of eighteen APCs. Two of them raced ahead to the next burning barricade at Xidan, fired tear gas there at 11:45 to clear demonstrators, but were forced to retreat and rejoin the column. Further back, there was fierce fighting at Fuxingmen, at a cost of some thirty civilian casualties.

Between the Xidan and Fuxingmen intersections on the north side of Changan stands the Nationalities (Minzu) Hotel. Directly in front of the hotel, a standoff between a riot squad and citizens grew into a pitched battle with over ten thousand men, women, and children at midnight. A foreigner staying the hotel saw the conflict grow from his window. "I watched about twenty men in combat gear fire tear gas into the crowd. Several young people were beaten to the ground. This provoked counter-reaction. The demonstrators threw stones and armed themselves with sticks and metal bars. For about one hour there was heavy fighting in front of the hotel. There were wounded, and perhaps also dead people. The victims were carrried away. Ambulances came."

The riot squad also suffered, according to a foreign photographer who arrived on the scene just after midnight. "The soldiers ran out of tear gas and some people moved in to get at them. Students and other people in the crowd tried, without success, to stop these people from hitting the soldiers. One soldier was hit with rocks and bars. He fell down but those hitting him continued. Several other soldiers were attacked in the same way. Some wounded or dead soldiers were taken into the Nationalities Hotel." The photographer followed them into the hotel for his own safety. "It was full of security people. When I tried to take a picture of the wounded soldiers, they attacked me. One of them tried to strangle me with the strap of my camera."

Eventually, the riot squad accompanying the soldiers in the street was able to form two lines along the sides of the boulevard to secure a traffic lane. Then at 1:00 A.M., about twenty or thirty APCs came from the west heading a great convoy of trucks loaded with armed soldiers. They moved in the direction of Tiananmen Square. The APCs simply flattened the street barricades. When after more than one hour the last truck of the convoy had passed by the Nationalities Hotel, where a foreigner was watching from a hotel window, many hundreds of people (not only students) appeared on the street. "They ran after the trucks and shouted protest slogans. A few stones were thrown. The soldiers opened fire with live ammunition. The crowd threw themselves on the ground, but quickly followed the convoy again. The more shots were fired, the more the crowd got determined and outraged. Suddenly they started singing *The Interna-*

tionale." Civilians did not confine themselves to throwing stones. "There were also a few Molotov cocktails, and the last truck was set on fire. The shooting continued. At first the soldiers shot over the heads of the people. Later the shooting continued in all directions. There were bullet holes in our hotel lobby."

The APCs at the head of this column crossed the burning barricade at Xidan and reached the northwest corner of the Square. A Chinese student was at Xidan when the first phalanx of infantry arrived behind them. The crowd there joined hands and yelled at them to go back. The troops paused for ten minutes. "Then, without warning, the troops opened fire on us. People cursed, screamed and ran. In no time, seventy or eighty people had collapsed all around me. Blood spattered all over, staining my clothes." Crouching at the side of the road as bullets streaked over his head, he watched several soldiers take aim at a limping girl and shoot her in the back of her white dress.

He and some friends followed in the wake of this contingent from Xidan to Liubukou.

> We saw bodies scattered all along the road. I must have seen several hundred bodies, mostly young people, and including some children.
>
> As the army reached Liubukou, an angry crowd of over ten thousand surged forward to surround the troops. This time the soldiers turned on the people with even greater brutality. The fusillades from machine-guns were loud and clear. Because some of the bullets used were of the kind that explode within the body, when they struck, the victims' intestines and brains spilled out. I saw four or five such bodies. They looked like disembowelled animal carcasses.
>
> I recall one scene clearly. A man with a Chinese journalist's identity badge on his shirt, waving a journalist's identity card all covered with blood, rushed toward the troops screaming, "Kill me! Kill me! You've already killed three of my colleagues!" Then I saw them shoot him and when he fell, several soldiers rushed over to kick him and to slash at him with their bayonets.

Another Chinese eyewitness arrived at Liubukou about an hour later. There was blood on the ground.

> I was told that a mass of people had blocked the army for a short time and were then crushed by tanks and armored cars travelling at full speed. Now only the last few trucks were surrounded by a group of people smaller in number than the troops passing through. Yet people were desperately shouting: "Lay down your arms, lay down your arms!" Every few seconds a soldier fired a clip of ammunition into the air. The people totally ignored him, except to lower their heads when he shot. Later, some people climbed aboard a truck and moved to disarm the soldiers.
>
> Suddenly about twenty-five soldiers from the next truck jumped down and opened fire. It all happened in a second. People fled in all directions. All

I could do was hide behind a rather thin tree at the side of the Boulevard. The fusillade of more than twenty guns continued for some eight minutes, some of the bullets splashing blue light on the ground. I saw seven people being taken to hospital on tricycles or whatever else was at hand. One man was seriously wounded in the abdomen, and several other hands plus his own could not stem the heavy bleeding. Some people rushed towards the trucks to save someone who had just fallen there, but the soldiers, thank heaven, didn't shoot at them. Once it was clear that no-one would oppose them, the soldiers helped those who had been besieged into their own truck and drove away at full speed, leaving behind the one with a broken tire. It was now about 3:30 A.M. I learned later that the soldiers on that truck (the third one from the end) lost their company commander.

The slain company commander was twenty-five-year-old Second Lieutenant Liu Guogeng, later placed on the PLA's "martyrs" list. According to the official account of his martyrdom, Liu's unit was surrounded on Changan Boulevard shortly after 3:00. Protesters disabled several of the troop and ammunition transports at the end of the column. When Liu realized that the tail of the column was not following the head, he returned west on foot with his sergeant to extricate them. The two men took off their Army jackets and left their rifles behind to avoid detection, though Liu did carry a revolver with sixteen shells. About 4:00 they were discovered at Liubukou. A crowd of people held and beat them for an hour. Liu escaped west, but other demonstrators captured him again, this time setting him on fire. He died about 5:30, and his corpse was strung up on the side of a bus at Xidan and torched. That is the official account.

According to the story on the street, Liu Guogeng was carrying his AK-47, and when he found himself surrounded, opened fire at point-blank range. When he ran out of ammunition, he was lynched. The government has published a gruesome close-up photograph of his burned body tied to the side of a bus. The photograph has been cropped to exclude a graffito someone scrawled on the side of the bus declaring that this soldier was "a murderer—killed four people, including a child."

The six-kilometer stretch of road from Liubukou back to the Military Museum was one of the toughest battlegrounds that night. The Army pushed, and the people resisted; the Army fired, and the people dodged the bullets but came right back again. The Army had to measure its progress in paces, not kilometers. It took the 38th Group Army, the main unit involved in this operation, four hours to cover the distance. The losses sustained by the 38th tell a tale of tremendous opposition. According to official statistics given later by Li Zhiyuan, chief political commissar of the 38th, 6 soldiers died and 1,114 were wounded. Sixty-five trucks and 47 APCs were destroyed; another 485 vehicles were damaged.

Civilian casualties along this stretch of road were also high. Commissar Li's estimate—over two hundred—undercounts the real toll. So too, his insistence that civilian deaths were the result of ricochets from soldiers' guns rather than direct fire is not borne out by other testimony. Bullet damage along Changan Boulevard confirms that soldiers were firing their guns at chest level. Even when guns were fired into the air above the heads of the demonstrators, as the Army insisted they were, the bullets took their toll. Large residential apartment buildings face down onto West Changan along both sides of the street west of Fuxingmen, among them some of the most exclusive residences in the city. The windows and balconies of high-ranking cadres gave them ringside seats to the carnage below. It also left them vulnerable when soldiers sprayed the sides of the large buildings with random fire or marksmen targeted lighted windows. Many are the stories circulating among Beijing residents of innocent people's being shot in their apartments that night as they sat at kitchen tables or brushed their teeth.

By 1:00 A.M., soldiers were starting to surround the Square in significant numbers (see Map 2). Those coming from the south were the first to come into position. They arrived at the southwest corner of the Square about midnight. An Army regiment that had set out from Marco Polo Bridge before 10:00 found its way blocked at the train tracks on the southwest side of the city. A freight train had pulled out of the trainyard to the north and blocked the road, forcing the soldiers to clamber under and between railway cars to get through. They were slowed down again in a street fight on the avenue south of Qianmen at 11:45, but managed to make it to the edge of the Square just before midnight. According to an official account, 80 percent of the regiment arrived wounded, 149 seriously. Shortly thereafter, the regiment was joined by a contingent of five hundred troops, alleged to be Air Force, also coming from the southwest but by a more northerly route. They too had been obliged to march on the double through a hail of bricks. Finally, a third contingent of soldiers marched to the south end from the Railway Station three kilometers to the east, where they had been holed up since May 21. Stragglers from that unit were still showing up at the Square at 12:30. Other units were also expected to be in position at the south end of the Square by midnight; those still bogged down began receiving authorization to go ahead at any cost at 12:05. Yet other units coming from the south had later deadlines. One colonel who ordered his men to fire into the air to disperse crowds declares he reached the south end of the Square at 1:25, which he says was "ahead of schedule."

People at the Square greeted the arrival of soldiers with anger. "Those of us on the Square had been hearing news that people were being killed, that the soldiers coming in from the west were shooting," recalled a Chinese eyewitness. "Some said that great numbers of people had already died. We didn't know how many actually had been killed, but we knew for certain that people had died. When the soldiers came toward the Square [at midnight], people were fired up, beyond reason. It was still unimaginable

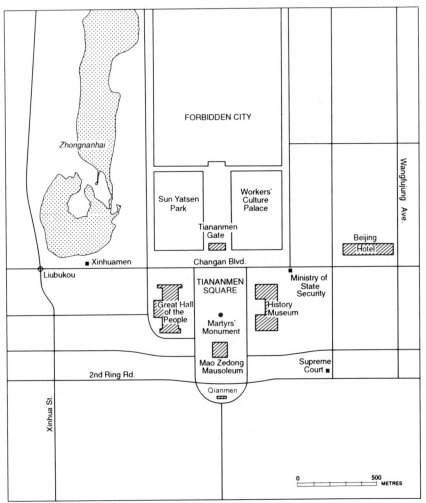

Map 2. Central Beijing.

to us that the government would shoot unarmed students and civilians with guns. Some young guys, city residents, started throwing bricks at them. At first the soldiers threw the bricks back at them. Soon after that I heard them open fire." The long volley echoed across the Square at about 12:30. "We did not run away. We were too amazed. The night was really black, so from where I was standing I couldn't see from where or at what angle they were. I kept down in a crouch position because I was afraid of getting shot, but a lot of people were standing, determined to see what was going on. I did see two people helping a girl in the crowd who got shot in the shoulder." Although the gunfire was not aimed at the demonstrators, there were casualties.

With the Air Force regiment had come two APCs. These vehicles had driven an opening for the troops through the blockade at West Qianmen Street, then proceeded to the southwest corner of the Square. They did not stay there, probably for fear of being attacked. Stangely enough, they repeated the bizarre episode of the two runaway APCs an hour earlier. The first drove up the west side of the Square at 12:15 A.M., turned east onto Changan, and went past the Beijing Hotel. The second drove around the Square some twenty minutes later and headed east past the Beijing Hotel at 12:50 A.M. Then both vehicles returned along Changan to the angry multitudes in the Square.

The first drove down the east side, uncertain of where to go but desperate to shake the crowds it was drawing. Gasoline from Molotov cocktails burned across its top surface, but it managed to escape and head off west. BBC journalist John Simpson guesses that it may have knocked down six or seven people. Then the second arrived, zigzagging its way through the barriers that people threw in its path. Unluckily, its tread got caught on a block of concrete. A mob descended on it, pouring gasoline over the metal armor and beating at the glass. Flames went up. It was 1:18. This APC was not ferrying soldiers. It carried only its three crew members: a commander, a driver, and a radio operator. Locked in the crowd a few feet from the vehicle, Simpson saw them emerge.

The screaming around me rose even louder: the handle of the door at the rear of the vehicle had turned a little, and the door began to open. A soldier pushed the barrel of a gun out, but it was snatched from his hands, and then everyone started grabbing his arms, pulling and wrenching until finally he came free, and then he was gone: I saw the arms of the mob flailing, raised above their heads as they fought to get their blows in. He was dead within seconds, and his body was dragged away in triumph. A second soldier showed his head through the door and was then immediately pulled out by his hair and ears and the skin on his face. This soldier I could see: his eyes were rolling, and his mouth was open, and he was covered with blood where the skin had been ripped off. Only his eyes remained—white and clear—but then someone was trying to get them as well, and someone else began beating his skull until the skull came apart, and there was blood all over the ground, and his brains, and still they kept on beating and beating what was left.

Then the horrible sight passed away, and the ground was wet where he had been.

There was a third soldier inside. I could see his face in the light of the flames, and some of the crowd could too. They pulled him out, screaming, wild at having missed killing the other soldiers. It was his blood they wanted, I was certain, it was to feel the blood running over their hands. Their mouths were open and panting, like dogs, and their eyes were expressionless. They were shouting that the soldier they were about to kill wasn't human, that he was just a thing, an object, which had to be destroyed. And all the time the noise and the heat and the stench of oil burning on hot metal beat at us, overwhelming our senses, deadening them.

Desperate to stop the killing, student monitors roared up in a public bus to rescue the soldiers the crowd had seized. The rear door banged open and a student leapt out to collect the only one still living, but the crowd would not give their victim up.

I saw the soldier's face, expressing only horror and pain as he sank under the blows of the people around him, and I started to move forward. The ferocity of the crowd had entered me, but I felt it was the crowd that was the animal, that it wasn't properly human. The soldier had sunk down to the ground, and a man was trying to break his skull with a half-brick, bringing it down with full force. I screamed obscenities at the man—stupid obscenities, as no one except my colleagues could have understood them—and threw myself at him, catching him with his arm up, poised for another blow. He looked at me blankly, and his thin arm went limp in my grasp. I stopped shouting. He relaxed his grip on the brick, and I threw it under the bus. It felt wet. A little room had been created around the soldier, and the student who had tried to rescue him before could now get at him. The rest of the mob hadn't given up, but the students were able to pull the soldier away and get him on to the bus by the other door. He was safe.

As flames rose from this APC, the detachments coming eastward along West Changan Boulevard were massing at the north end of the Great Hall of the People, just beyond the Square. A young woman grabbed a foreign journalist over at the northeast corner of the Square and told her that soldiers were shooting people by the Great Hall. The journalist followed her toward the northwest corner just as the soldiers opened fire across the top of the Square at 1:13. "I could see soldiers and APCs, and I could hear a lot of gunfire. The firing was coming from farther west." The APCs on Changan behind the soldiers fired tracer rounds from their machine guns over the Square. Not wanting to be caught in a hail of bullets, she retreated east just as the captured APC was set alight. The north end of the Square, beyond the burning APC, was about to become a battleground between soldiers with guns and young people with nothing but voices.

The firing that started at 1:13 was indiscriminate, and direct. Those who stood at the northwest corner of the Square expected no worse than rubber bullets which are fired to ricochet off the ground and strike the legs of fleeing rioters. Assuming that this was what the soldiers would use, the students and young workers "held up coats and quilts to protect themselves from the rubber bullets when they were shot at. But some fell backwards and didn't come up again. Students went forward and fetched some of them, the wounded and the bodies, moved them to the side."

Rather than walk into the soldiers' onslaught, the journalist decided to go into the main part of the Square to reconnoiter the situation. "Almost the first person I bumped into was a Beida student I knew. I begged him to leave, and he begged me to leave. He said, no, he wouldn't leave, the students have a plan. They were going to link arms and all gather in circles around the base of the Monument, and they would just sit or lie there together and accept whatever was coming their way." This is what most of the students on the Square did. "Another student I spoke with, from Sichuan, told me their plan was to lie down in their tents. His tent was near the front of the area where the tents had been lined up. Maybe fear got him in the end and he left, but he wanted to remain in his tent. Other students from Sichuan had exactly the same intention. I begged him to go but he wouldn't leave."

Most of the students chose to wait quietly near the base of the Monument, armed only with damp clothes against the tear gas. Some hotheads started yelling for bricks, though as of 1:30 there were still no fighting and no casualties within the perimeter of the Square itself. Someone looked into the medical tents by the Monument just at this time and found two young men lying on the ground, probably from heat prostration, exhausted. There were no shooting victims yet within the Square.

Meanwhile, the government loudspeakers, which at 1:10 had warned people to leave the Square, were now broadcasting a continuous warning that anyone caught there would be considered as participating in a counterrevolutionary rebellion. The Army, it said, had been attacked by people committed to overthrowing the Communist Party and would now no longer exercise "restraint." It warned that the personal safety of those who disregarded this warning "could not be guaranteed." They alone would be "held responsible for all consequences." As journalist Jan Wong noted, this sort of announcement had been made before, and the threatened "consequences" had never materialized. "The broadcast told us to go home and that we weren't supposed to be out in the streets, but it didn't say: We are going to kill you. It didn't say: Soldiers are coming and are going to shoot to kill. It just said that everybody should go home. The government had lost all its credibility. No-one paid the slightest attention to what was being said."

At about 1:40 A.M., the battle for the north end of the Square began in earnest, launched from the northwest. Jan Wong was near the north end of the Square when she found herself in a stampede. "Everyone panicked and ran without knowing why. The only place I could stand and take refuge was behind a lamppost. And there were already thirty other people trying to take refuge there. I heard people say soldiers were entering the Square. I didn't think they would shoot but I didn't want to be around when the people really panicked." She returned to the Beijing Hotel, where a special PSB team had just emerged to frisk foreign reporters and confiscate films and notebooks. (Hotel security cut the telephone lines—literally, with scissors—at 3:00.) She positioned herself on a balcony facing the street and began to make a careful record of what happened at the eastern approach to the Square for the rest of the night. (I have relied in part on her diary to write the account that follows.)

Just before Wong went into the hotel, CBC journalist Tom Kennedy headed out toward the Square to the sound of periodic gunfire west and south. The closer he got to Tiananmen, the more nervous the crowd became. "I got to the northeast corner of the Square and could see soldiers firing, mostly into the south and southwest corner. An APC was in flames at the north end of the Square right in front of the Forbidden City. The light from the flames reflected off their helmets. They walked slowly as they fired, and their rifles were on semi-automatic. They weren't firing into the ground or above people's heads—it was straight." Kennedy took cover behind lampposts at the northeast corner of the Square and watched as people were carried out of the north end with head and chest wounds. "I could see over in front of the Great Hall of the People, not clearly because it was quite dark, but there was a lot of movement over there, soldiers." By this time the tent serving as the headquarters of the Workers' Autonomous Union in that corner of the Square was in flames.

Another CBC journalist was also in the crowd at the north end of the Square when the Army began to push from the west. She was taken to see some of the casualties at close range. "Can you come? Can you come?" a woman and two men cried, running up to her. "There are bodies."

They led her to the north side of Changan, almost directly beneath the portrait of Mao that hangs over the main tunnel through the gate of Tiananmen. Four bodies lay there, stretched on the ground. They had been lined up beside each other, as though dragged there from wherever they had been shot. One had a bullet through the neck, another a bullet between the eyes, another a gaping wound in the middle of the chest. Every one of them had been shot from the front.

The firing suddenly started up again. "It was really close. I'd say some of it was almost to the gates of the Forbidden City. A man came running up and pulled his jacket open: his shoulder had been blown away. Then the crowd parted like the sea and a girl who'd been shot right between the eyes was brought through. There wasn't much left of her face. Not even two minutes after she was dragged out, a man was brought out. He had no head."

For an hour she watched casualties being removed. "I saw bodies on pedicabs, slung over bicycles, on flatbeds piled on top of each other. People were also bringing the wounded over the little bridges in front of the Forbidden City and into the street, as though they'd been shot from the palace walls to the north. And there were firecrackers being let off from the walls of the Forbidden City. They were as frightening as the gunfire, adding to the noise." Desperate to escape from the battle, at least one man jumped into the old moat around the Forbidden City and swam to safety. Several dozen fled into the Forbidden City itself when the gate at Tiananmen opened, though as it did so a large detachment of Armed Police emerged. Even larger numbers of PAP were waiting inside the palace grounds.

Many spectators still could not grasp that the situation had changed: the soldiers were not simply driving people off the streets but shooting to kill with live ammunition. As late as 2:10 A.M., a man insisted to a foreign eyewitness he was speaking with that even if the soldiers did fire, they would be using rubber bullets only. The truth would hit home shortly to all in Tiananmen as the forces coming in along West Changan Boulevard arrived in large numbers at the northwest corner of the Square. A picket at that corner watched several thousand riot police equipped with helmets and transparent shields advance under a volley of gunfire shortly after 2:00 A.M.

> The students lit a fire barricade across Changan Boulevard in order to block their advance. At the western entrance to the street underpass to Tiananmen, a picket suddenly collapsed less than a meter away from me. Blood gushed from a bullet hole in his neck three centimeters in diameter. I was stunned. By the time I got him to the emergency medical station, there was no sign of life in him. I found ten to twenty other people there, with bullet wounds to the stomach, chest, and head. The doctors there told me that people were definitely being shot with deadly military bullets, not rubber bullets.
>
> We retreated to the stretch of Changan Boulevard directly across from the gate of Tiananmen. Because many students still did not know that the military had opened fire with real bullets, no-one went for cover and casualties were high. Next to me, five or six students went down. One student who had been wounded in the leg crawled over to the side of the road to dodge the soldiers' billy clubs, trailing a track of blood behind him.

A Capital Iron and Steel worker who witnessed the assault that drove protesters back from the northwest corner of the Square was able to hide from the line of fire and watch the soldiers advance. Between 2:10 to 3:03 he counted twenty-nine young men and women shot dead.

The Army's objective at this point was not to take the Square but to clear everyone off Changan Boulevard at the north end and establish a cordon to prevent anyone outside the Square from joining those already there. A foreign journalist at the northeast corner at 2:15 A.M. watched the bodies of those who had fallen during this encounter being brought out. "Every few minutes, there was a burst of gunfire directed toward the crowds. People dispersed, hit the deck, ran for the trees along the side of the street, but then they would quickly regroup and head back into the street. A constant stream of bodies came out of the Square by ambulance and pedicab. I saw one chap being carried on the back of a bicycle toward Capital Hospital. Where his leg used to be there was nothing but a bloody stump."

By 2:35 A.M., the soldiers had established their cordon across Changan Boulevard just beyond the northeast corner of the Square. They would remain there for the rest of the night and all next day, drawn up in two lines, shooting to kill whenever they judged the crowd was getting too close. The first time they opened fire was at 2:40 A.M. "People behind us were pressing us toward the Square," recalls an eyewitness in the crowd. "All of a sudden, a white wall opened up right in front of us. You could see the guns firing, and people ran like crazy. Bullets ricocheted off the walls, which is what I was afraid of, because I couldn't believe they were shooting right at us. Some people went down. But after a while we all moved right back. It was like a dream. You are being shot at, and you can't believe you'd be so stupid as to go back again, but you do. I knew that if I left, I would witness nothing, know nothing of what really happened." He remained where he was, thousands of others at his side.

"The crowd started singing *The Internationale*, but not the way the students sang it. That never seemed authentic to me. These ordinary people were really singing it. They just got past the first verse when another wall of gunfire opened up in front of us for the second time." The time was now 3:12 A.M.

A Chinese artist who had arrived a few minutes earlier found herself caught in the action. "People were becoming more and more enraged, calling the soldiers animals. Some of them joined hands and walked toward the soldiers, singing. I was one of the them. Somehow at that moment I didn't feel any danger. More and more people joined us as we advanced. I realized that I was almost at the forefront of this group of people. It was terrifying, what we did, looking back on it, but it didn't seem so at the time." Then 3:12 struck. "Suddenly, the soldiers began to fire into the crowd. Everyone turned and ran. I fell, tripping over a bicycle on the curb. I heard bullets flying over my back and thought, people will trample me. But no-one did. The shooting stopped; I stood up. People around me were crying out that they were wounded. Others were carrying them back in carts or on bicycles. The ambulances were still unable to get in."

A doctor in a white coat was among those who dropped to the ground when the firing started. He stood up now and walked forward to where the injured lay.

"I need to get in to rescue the wounded!" he yelled to the soldiers. They paid no attention to his call and shot him. People were furious. They gave up "The Internationale" and switched to an angry chant.

"Down with Deng Xiaoping!" they cried. "Down with Deng Xiaoping!"

As the northeast exit from the Square was closed off, some trapped inside the cordon sought other avenues of escape. One student on the west side took off around the back of the Great Hall of the People and hoped to head west along Changan Boulevard. At that moment, the Army was sweeping the street with gunfire at Liubukou. A bullet broke his leg. He was left to bleed to death.

Two Canadians who also found themselves on the wrong side of the cordon at the northeast corner ended up traveling halfway across the south city that night to escape the Army. They had arrived at the Square at 1:30 A.M. They walked toward the two APCs burning at the north end, but got only as far as the History Museum, which flanks the east side of the Square, when they heard the sound of gunfire. People around them insisted it wasn't live ammunition. "Then some people started screaming that they were hit by bullets, that there was live fire." This was the beginning of the push to close the Square. The crowd stampeded, and S. and J. decided to run for cover to the stairs at the north end of the History Museum. "There was gunfire. Everyone screamed. We hit the dirt at the side of the street, crawled to the hedge and dove over the bushes. We made our way on our elbows over to the stairs and got to a solid concrete stairwall that separated us from the Square. Half a dozen Chinese were already sitting there." The stairwall protected the group from any fire coming from the west, as well as from the east for that matter, and they had the History Museum at their backs. They were exposed only from the north: Changan Boulevard, and beyond that the east wing of the Forbidden City. A marksman crouching on the palace wall could have picked them out clearly against the white marble of the steps.

J. felt something whiz under one leg, a single bullet rather than the uncontrolled firing of an AK-47. "I think someone's shooting at us," J. said to S. "I just felt a bullet go past my leg." S. thought he was making a joke, but before he could reply, a second bullet shot past him and struck the man sitting beside him. Their illusion of safety was shattered.

"We dragged him up the steps behind the three-foot wall onto the plinth supporting the pillars around the museum. His brother and about four other Chinese were with us. Here we were safe from fire from all directions. We got the guy's pants down and saw the wound was nasty. Two bullets had gone into his groin right next to his genitals. J. ripped off his shirt and tied tourniquets around his legs. He was in shock, throwing up."

The second retreat to safety behind the Museum pillars also proved to be illusory. "I was holding his hand, and the next thing I know the entire steps had filled with several thousand soldiers carrying iron bars. They didn't come from in front of us, but from the Ministry of State Security east of the museum. They had a wild, agitated look in their eyes, like animals about to stampede. They looked at us like we were just objects that had to be smashed or killed. They were in an extreme state of agitation. Maybe they thought they were going to die. It seemed that the slightest thing would trigger them to kill us."

Everyone froze. "Don't say anything to them," J. whispered to the others. Then S. decided that perhaps the sound of a human voice might break the spell and save their lives.

"I am a Canadian," he said carefully in Chinese, "and this man is injured." It crossed his mind to say something about Dr. Norman Bethune, "but then I thought that's just enough to end our lives. These soldiers didn't seem to be in their right senses at all." There was no response. He tried again.

"Can you tell us where we can go to get a doctor?" Again no response. Suddenly an officer gave a command. A few of the soldiers ran behind a nearby pillar where sharp wooden hexagons used as barriers in crowd control were stacked. They grabbed them and put them across the plinth between them and the group of terrified civilians, as if to cut off an attack from the west side of the museum. "This must have been planned, since the hexagons were up against the wall waiting to be used for the defense of the museum. The soldiers all turned to face us right behind these things, with their bars up. I didn't feel that they were going to attack us: they thought *they* were going to be attacked, right down this corridor where we were lying. I thought, shit, we don't know what's happening around the corner. We're going to die in the middle of someone else's battle." He tried a third time.

"This gentleman is wounded. Can we take him over behind this pillar?" Nothing.

"Don't say anything," J. told the Chinese with them. "Let's just drag our friend here over to the pillar." Slowly, they inched the wounded man back behind the west side of a pillar, without disappearing completely form the soldiers' view. The squad of soldiers turned north to face the crowds below.

"Get together! Get together!" shouted their commander. They massed in a tight formation and raised their clubs, poised. Then the litany began.

"Protect China!" barked the commander.

"Protect China!" the soldiers yelled back.

"Protect the capital!" he yelled above the din.

"Protect the capital!" they chanted back.

"Protect the people!" was his third cry.

"Protect the people!" they replied. On that signal, they charged into the stampeding crowd on the street below, bringing their iron bars down on the backs and heads of people as they tried to flee.

The crowd was defiant in flight. "On strike!" they yelled back. "On strike!"

The soldiers regrouped and charged again. The workers ran, but as they ran they kept yelling, "On strike! On strike!"

The small group huddled by the pillar watched in horror at the scene below. "After those two charges, a worker jumped up on the wall seven or eight meters from us. He had an iron bar like theirs, which he must have pulled from the hands of a soldier. I couldn't see what he was planning to do. The soldiers had already regrouped. Then one soldier ran forward and they both brought their bars up. The soldier was a little bit faster and won, striking him over the head with tremendous force. The man came down like a ton of bricks and dropped into the bushes beyond the wall."

As the soldiers awaited the order for the next charge, some were muttering, "Foreigners, foreigners." An officer came over where they lay. "You can pull back," he said.

They pulled back about five meters closer to the inner wall and lay there for a while. Then the officer came back and said they could leave.

"We stood up and started to carry the wounded guy, when the troops rushed up with clubs raised and surrounded us. They hadn't heard their officer's order that we could move. We were down on our knees with our hands above our heads saying, 'Spare us! Spare us!' Just in the nick of time, the officer came back."

"They can go," he barked, taking the moment of death away. "We grabbed the injured man's arms and legs and started to run. They pursued us, just to make sure we ran. When we got around the corner onto the west side of the Museum, which faces the Square, the plinth was empty. But the wide set of steps opposite the Monument was completely filled with another huge body of armed soldiers, a couple of thousand, who were peacefully sitting there like Buddhas. You couldn't hear any of the noise to the north. Some students were talking to soldiers on the steps."

An officer approached them immediately.

"Is he a student or a worker?" he demanded.

"He's a worker," they replied.

The officer looked over the man's clothes to verify this identity, then let them proceed. One of them ran over to the Red Cross station on the Square for a doctor, surprised to see no signs of the fighting they had just witnessed. None of the casualties had been brought down into the Square. Doctors went to fetch the man.

As the two foreigners turned to go, one of the doctors said, in English, "Thank you very much."

Their first thought was to stay and see what would happen. Their second was to escape. The only direction that seemed to be open was to the southeast by Mao's Mausoleum. They walked in that direction. Once out of the Square, they discovered troops approaching the south end of the Square along both West and East Qianmen Streets (East Qianmen connects the Square with the Railway Station). Their only choice now was to go due south on Qianmen Street, but they got no further than the next major intersection when they saw a military convoy coming north toward them. It was just before 3:00 A.M. They ducked into an alley in the hope of working their way eastward to the next major road, but local residents stopped them.

"Where do you think you're going?" someone at the head of the alley said. "They won't come in the alleys, because this is where the people live. They might try to hunt us here in the daytime, but they know they could get killed at night. Anyway, we're in our homes, we're supposed to be here." He led the two into their homes and listened to their account of what they had seen at the Square. Then they all went out to see what was happening along Qianmen Street. The alley gave the residents and their guests a safe roost from which to observe the Army's fierce northward advance from the south end of the city.

The first armored vehicles had broken through the southern round-about on the 3rd Ring Road about midnight. An hour later, a column of 120 troop trucks left Nanyuan Airfield, crossed the ring road, but was stopped well down Qianmen Street, unbeknownst to the students in the Square. Supporting units found their way blocked on other roads as well. Further east, at the east gate of the Temple of Heaven, a regiment of 880 men coming from Nanyuan was halted at 2:00 A.M. by thousands of people. Under orders to be in position southeast of the Square at 4:30, the regiment fired into the air around 3:00. Other units were doing the same. The south end of the city would echo with gunfire and explosions for the next three and a half hours. An eyewitness who passed by the Temple of Heaven the next morning described the damage of the night before: "It was as though a civil war had taken place. Everywhere, every intersection, there were burned-out vehicles: jeeps, military vehicles, APCs. We saw soldiers with machine guns on guard inside the Temple of Heaven. Out-side the north entrance, five military vehicles were piled up on each other, completely burned out."

The main forces from the south, however, came directly up Qianmen Street. From the mouth of their alley, S. and J. watched. "There were so many people out, but the soldiers came through shooting, firing directly ahead to clear a path. People got in the way. Six were shot at the nearby intersection early on and were dragged off into the alley across from us. One was obviously dead, the other five dying. I was surprised how brave—or incautious—the people of the south city were that night. They knew people were being shot and yet they ran to the intersection to watch for the next wave of troops coming."

At 5:00 A.M., a man on a bicycle stopped at the end of the alley where they and several others had posted themselves. He passed around a large shell he had picked up off the street. He was furious.

"What do these people think they're doing?" he screamed.

One of them measured the shell in his hand. It went from the base of his thumb to the end of his fingers. His measurement suggests a 14.5 caliber, the size of shell fired from the Type 75-1 antiaircraft machine gun mounted on APCs. The man soon sped off on his bicycle to show it to others.

Through the predawn hours, wave after wave of troops walking briskly in columns about five meters wide, sometimes accompanied by

APCs, pushed north up Qianmen Street to join the forces surrounding the Square. "The last time I saw soldiers go by—and this troop movement lasted for a long time—was after the street lights went out," which was at 4:30 A.M. As the soldiers marched north, the people rained abuse on them.

"Yi, er, faxisi!" they chanted. "One, two, fascists!"

"From alleyways, from doorways, from wherever they were, everyone yelled the same thing. I wondered what the soldiers were thinking as they walked into the city and heard nothing but the unbroken chant of the people yelling, 'Fascists!'"

Civilians were not alone in suffering casualties in the south city. The best-publicized incident involving the killing of a PLA soldier occurred an hour before dawn at Chongwenmen, the intersection at the southeast corner of the 2nd Ring Road where it curves up to the Railway Station. Chongwenmen is the main intersection on the east-west route from the Railway Station to the south end of the Square, as well as on the north-south route linking the Temple of Heaven to the Dongdan intersection on Changan. The intersection was blocked with buses by midnight, though detachments of foot soldiers coming from the station were still able to get through on their way to the Square.

According to State Council spokesman Yuan Mu at a press conference on June 6, Cui Guozheng was one of eleven soldiers attacked by a mob for no reason at Chongwenmen. The others escaped, but Cui had the misfortune to be caught, thrown from the Chongwenmen pedestrian footbridge, then drenched with gasoline and set on fire. His charred corpse was slung from the footbridge and left for all to see until soldiers came to remove it Sunday afternoon. He was blameless, killed by rioters without a gun in his hand. Two foreign eyewitnesses in the Hadamen Hotel on the southeast corner of the Chongwenwen intersection have given Amnesty International a different account of the killing of Cui Guozheng.

Sometime after 3:00 A.M., three Army trucks came from the south, their backs covered with canvas. The crowd, which was then gathered at the intersection, moved south from it toward the trucks, which they surrounded. While this happened, a group of several hundred soldiers marching in disciplined formation came from the east, crossed the intersection, and stopped on the west side. They formed two rows facing east, some kneeling, some standing, and fired toward the east for several minutes. From their position in the hotel, the eyewitnesses could not see whether anyone was hurt on the east side of the intersection. After firing, the troops marched west toward Tiananmen Square.

Meanwhile, two of the trucks in the street to the south had turned around and gone away. The third, which had a trailer, tried to turn around but was stuck on a piece of pavement. The crowd pelted the front of the truck with bottles. Soldiers in the cab tried to get out. Two eventually did. One, wearing a helmet, was approached by three men. He pulled the hair of one and was set upon by people around him. He then disappeared. The other soldier went back into the cab and re-emerged at the back, holding a rifle. He fired at the crowd. The eyewitnesses heard three shots and saw an old woman and a man fall. They later heard that a child had also been shot. The crowd was incensed and stormed into the truck. The soldier re-emerged from the cab, his clothes half-torn. He ran toward Chongwenmen Hotel (on the west side), managed to reach the sidewalk, but was dragged away to the left. The eyewitnesses could not see clearly what happened to him next—the pedestrian footbridge blocked their view—but they assume he was killed. The next thing they saw was a fire burning. Later in the morning, they saw the soldier's charred body hanging from a rope on the pedestrian footbridge.

After the soldier was killed, the crowd kept at some distance from the truck. Gradually, some rifles emerged from the back of the truck, but no-one came out of it for about half an hour. Some people in the crowd threw gasoline bombs toward the truck. The truck eventually caught fire and soldiers got out of it. They ran over the footbridge, holding their rifles toward the crowd. The crowd, still numbering several thousand, ran toward the east side of the road. The soldiers did not shoot but ran backward up to a side street, into which they disappeared.

The killing of Cui Guozheng was an atrocity that supporters of the Democracy Movement can only regret. But the responsibility for such evil does not rest solely with the crowd that attacked him. It also lies with Cui for opening fire on the crowds; with his commander for failing to take control of the situation; with his commander's superiors for sending the trucks toward Chongwenmen without sufficient protection; and with the Party for ordering the violence to begin.

The noose was tightening. Soldiers in the tens of thousands were now pouring through Chongwenmen and other intersections toward the Army's final goal, Tiananmen Square. At 2:40 A.M., fifty-five hundred troops came from the Railway Station via Chongwenmen and entered the Square from the southeast. Armed infantry entered from the southwest about twenty minutes later. Large numbers in battle dress emerged from the Great Hall and lined up along the west side of the Square. And troops continued to pour in from the west, filling the width of Changan Boulevard behind the cordon. Behind them followed a truck convoy carrying ten thousand soldiers, part of which consisted of soldiers from the 38th Group Army. The convoy was halted at the Fuxingmen overpass over the 2nd Ring Road, presumably to wait until the Square was ready for the taking. At 4:15 A.M., the column started up again, reaching the Square just before the clearing operation was to begin.

Those outside the Square, dodging the bullets of the soldiers at the northeast corner, heard incessant gunfire and feared the worst. "Almost constant automatic gunfire came from Tiananmen Square," recalled one journalist. "From the sound of it, we thought a slaughter was going on in the Square." In actual fact, the Square was not being bathed in blood, as everyone fancied. Neither, however, was it being occupied without the use of force. Soldiers did fire. At 3:45 A.M., just as a helicopter hovered over Tiananmen, someone phoned Reuters to report just that. The Chinese government has since denied that anyone was killed within the perimeter of the Square between 4:30 and 5:30. This denial, if true in some narrow, literal sense, applies only to the Square proper and not to the broad streets bounding the Square, and only for that one hour. It cannot dispute the fact that at many times on and around the Square Sunday morning, soldiers were killing civilians. All that eyewitnesses outside the Square knew was that dead and wounded were being brought out to them without a break during the two hours leading up to the final clearing.

Just before 4:00 A.M., Wuerkaixi was evacuated from the Square in a bullet-scarred ambulance. It was felt that his leadership in the Movement had jeopardized his personal safety and that he and other prominent activists should not remain in the Square when the Army took over. (Hou Dejian would be wheeled out later on a pedicab, hidden among the wounded.) Wuerkaixi testifies that students were shot on the west side of the Square just prior to his escape. "There were four casualties inside the ambulance. When we got to the hospital, two were already dead. One was lying on me. They were shot within the confines of the Square, in the area in front of the Great Hall of the People, sometime between 3:30 and when I got in the ambulance. They were hit by bullets not at close range but from quite a distance, fired by soldiers coming out of the Great Hall. The backs of their skulls were completely shattered. The brains of the one dripped out on my right side in the ambulance."

It became vividly clear to some on the Square, particularly Hou Dejian and the three who staged the final hunger strike on Friday, that the task now was not to embrace the futile gesture of sacrifice but to save as many lives as possible. According to Hou's own testimony, they set about trying to convince the other leaders to disarm the workers who still held weapons. Their cache included a captured machine gun, two automatic rifles, and a pistol, in addition to Molotov cocktails. Despite some resistance, all the weapons were finally dismantled or destroyed, and photographs taken to confirm that this had been done.

At the urging of two Red Cross doctors, Hou Dejian then led a delegation of four to negotiate the students' withdrawal. They met first with a regimental commander, then with two officers from the Martial Law Command. By telephone, the headquarters of the Beijing Military Region agreed that the students be allowed to withdraw voluntarily by the southeast corner of the Square. The withdrawal should be completed before dawn. The order forbidding the soldiers from opening fire was confirmed for Hou by Colonel Ji Xingguo, political commissar with the 27th Group Army.

At 4:00 A.M., the lights of the Square were turned off. A flash of anxiety swept over the students, but nothing happened. For the next forty minutes, the Square was draped in near-total darkness, except for the lights inside the Great Hall, where the PLA leadership had gathered. Cutting the lights was the soldiers' signal to get into position for the clearing operation. Soldiers at the northwest corner, joined at 4:15 by thousands from the Great Hall (some with fixed bayonets) and the Forbidden City, formed two lines, one facing out and the other facing in. Roughly a dozen APCs took up positions at the north end along Changan Boulevard. Three thousand soldiers squatted on the steps of the History Museum. Down at the south end, security forces in riot gear took up positions on the grounds of Mao's Mausoleum. The clearing operation would be a squeeze from four sides, leaving the one promised exit in the southeast corner. At 4:30, the commanders of the PLA emerged onto the roof of the Great Hall to watch the operation.

The majority of the students grouped around the base of the Monument wanted simply to remain where they were and take whatever was coming. They felt that to retreat was to surrender, to admit failure. Only total self-sacrifice would be worthy of the moment. At 4:15, the government broadcasting paused and someone got on the students' public-address system to tell the military to leave the Square. Then, ten minutes later, Friday's hunger strikers took over the microphone and surprised everyone by explaining that they had struck a deal with the Army to let the students evacuate. A first vote turned down their proposal.

At 4:40 a.m., the lights came back on. Sudden gunfire blasted away at the student loudspeakers on the Monument piping out "The Internationale." One student remembers seeing "sparks flying in all directions" from the top of the obelisk. Another standing at the base of the Monument recalls the moment the guns went off with disturbing clarity. "Immediately the troops surrounding the Square began firing indiscriminately. Stray shots flew around the Square. A girl who was about three meters away from me suddenly went down with a bullet in her head." He wrapped her head in a piece of clothing and with several others rushed her to the medical station on the east side of the Square but she was dead when they got there. He found four other gunshot victims lying there. Perhaps the bullet in the young woman's head was a stray; perhaps that is why the government felt it could claim on one was "killed" by a soldier in the hour between 4:30 and 5:30. The telltale signs of repaired bullet holes I found four months later on the west and south sides of the platforms and balustrades around the base of the Monument warn that she may not have been the only casualty.

Over the government loudspeakers came the final order: "Students, clear the Square immmediately!" Only now could the students see clearly that troops in camouflage had massed around the perimeter of the Square, and that the Goddess of Democracy had been toppled. (The soldiers assigned the task were supposed to bring the head back to the Martial Law Command but found it too firmly attached. They lopped off the torch instead.) Student leaders insisted that retreat was the only reasonable course left. The voice vote at 4:45 was ambiguous, but it was deemed to favor withdrawal and the students organized themselves to leave. Hou Dejian and others dragged the reluctant students to their feet to get them moving. The first began to file out at 4:55.

As the exodus started, the APCs lined up side by side at the north end of the Square began to roll south. Robin Munro of Asia Watch saw them coming:

> A row of armored vehicles was moving very slowly in a straight east-west formation at no more than two or three miles per hour from the top of the Square down toward the Monument. The line of armored vehicles was about twenty-five meters from where we stood, there was a gap of about two meters between each vehicle, through which was plainly visible a dense line of helmeted troops, following directly behind the armored line. The tanks and APCs were simply crushing everything in their path—tents, railings, boxes of provisions, bicycles—all were knocked over and squashed flat, in a slow-motion display of military might. The sight was mesmerizing in its sheer brutish idiocy and its random purposelessness. We stood there staring at it for about three minutes.

The accompanying infantry was from the 27th and 38th, and possibly other group armies. The soldiers' task was to ensure that the two hundred tents were empty before the APCs flattened them. A military spokesman two days later stated that the soldiers performed this task properly, that no one was crushed inside the tents when the APCs moved south. At the end of the week, Deng Xiaoping repeated the government's assertion, adding that "if tanks were used to roll over people, this would have created a confusion between right and wrong among the people nationwide." Deng was right about the negative impact that news of tanks rolling over students would have had: the mere *belief* that they did spread immediately through China and did much to blacken the government's reputation.

Evidence that students were flattened in their tents is only secondhand. Many had vowed to remain where they lay to the bitter end. Did they keep to that vow? We do not know. A Beida student did photograph what appeared to be the corpse of a crushed demonstrator in a tent. He displayed the picture at Beida at 7:30 that morning. "It was the most horrible photograph," as someone at Beida who saw it told me later. "It showed a student lying under a quilt in one of the tents on the Square. You could tell he was in the tent by the way the space was organized. A tread had run over him and split him in two, from above the knees to the abdomen. The guy who took the picture was walking around showing it to everybody. The picture was so clear. There was no doubt."

Some three thousand students walked out of the Square in the hour between 5:00 and 6:00. They marched by school, some still carrying their school banners. They had to march out of the southeast corner through a gauntlet of troops lined up in double columns on each side. The inner line of soldiers faced the students and pointed their guns at them; the outer line faced away and shot into the sky to terrify and humiliate the defeated students. "Many helmeted military police, most of them PLA, carrying billy clubs and electric cattle prods pressed in from all sides. The opening through which the students were leaving became narrow, and this increased the chaos. Students were trampled down and hurt; cries of pain rose constantly; many were wounded."

The Army appears to have been under orders to have the Square cleared by sunrise. At 5:23, when dawn broke, students were still slowly filing out of the southeast corner. The soldiers became increasingly anxious and aggressive, pointing their guns at the students and harassing them to hurry. Guns were also fired, starting at 5:04 and continuing intermittently for the next forty minutes. Most of this gunfire was warning shots aimed over the students' heads. Around 5:20, a tremendous volley of gunfire resounded from the southeast corner for over a minute. The Army later maintained that the soldiers were trying to flush out a sniper in a building overlooking the Square, though no sniper was ever captured. Popular rumor has it that the sniper was a member of the PAP.

Did any students die during the exodus from the Square? I have already noted that a military spokesman two days later insisted no one was killed on the Square between 4:30 and 5:30 A.M. Because the last of the students were not completely out of the Square for at least half an hour after that, it is possible that the statement, even if it is not an outright lie, was designed to obscure killing after 5:30. A student leaving the Square has testified that at least one teenage girl was severely injured in the southwest corner of the Square around 5:30. "She must have shouted something at a soldier. An armed policeman rushed toward her, and using a blade that was about a meter long slashed at her. In that instant she raised one of her arms to defend herself, and her arm was literally chopped in two. It dangled and flopped at her side, held on by strands of cloth and flesh. Blood spewed from the wound like a fountain. The girl screamed and screamed, her arm still dangling."

Other eyewitnesses report sharp bursts of gunfire at 5:40 and 5:45, not all of it from the soldiers on the exit corridor. But none has provided conclusive evidence of further killing in the Square. One student source insists that some students from Qinghua University refused to leave the Monument and were shot. Another says that students at the tail end of the file coming out the southeast were shot. Yet another holds that an APC opened fire on students who had retreated to a building near the south end of the Square. On the other hand, several student leaders and observers have gone on record to say that, at least in the southeast corner of the Square, they did not see the Army shoot any students during the final steps of the clearing operation. But what any one person can see is limited in so vast a space and so confused a setting.

After leaving the Square, some turned east along Qianmen East Street in the direction of the Railway Station. Robin Munro was with them when soldiers opened fire at 6:15.

> I noticed three PLA soldiers, their AK-47s levelled, come walking along the pavement directly across the road from where we stood. The small crowd around us then also spotted the soldiers, who were moving cautiously and guardedly as they came, and people started shouting and running toward them, throwing stones at the soldiers. The soldiers then opened fire, shooting wildly in brief bursts all around them, as they began to run off to the east along the southern side of Qianmen East Street, hotly pursued by the crowd.

Most of the students exiting from the Square headed west, however, walking in formation. As they passed a group of troops sitting at the southwest corner of the Square, students yelled insults. Again, the soldiers opened fire, though the volley went safely over their heads. No one was hit. The column of students continued west and then north to the Liubukou intersection, then turned west on Changan. As they rounded the corner, some of the students shouted insults at a group of tanks parked a hundred meters east of the intersection. Four of the tanks started up their engines and charged the students. Some demonstrators couldn't move quickly enough over the debris of burned-out buses and tangled bicycles. Tank 100 ploughed into the tail end of the students' column, pivoted 360 degrees, then raced back to its original position.

"I had to flatten myself against a wall," a student caught in the attack testified. "After the tank passed, I ran back and looked, and counted eleven bodies. You've seen cartoons in which steamrollers roll over people and they get flattened? That's exactly what happened in front of me." A Beishida student witnessed the same atrocity. "We didn't fight anyone, we just left. But they chased us with a tank, ran us down. I saw eleven students crushed alive when the tank roared up behind them. After they were crushed, they didn't look like humans." An Associated Press reporter also witnessed the attack: "Seven died instantly and four probably died later. They were like hamburger, like a dead animal flattened on the highway. Maybe the driver just lost control, though I assume it was on purpose."

The Beida student who photographed the severed body of a student in a tent also took pictures of this scene and displayed them that morning on the campus. "I had a hell of a problem identifying this photo because it was so horrible," recalls a foreigner who studied one of the pictures. "It was a body, but the head of the person had been smashed, run over by a vehicle, so that it was just paste. It was difficult to identify because it was such a mess, this body in the debris, and paste where the top of the body should have been."

The tank attack on the retreating column of students was the final grim gesture to the clearing of the Square. If it was the doing of one angry tank commander, it betrays a disturbing level of indiscipline within the PLA. If the charge was ordered by his commanding officer, it shows how great had grown the gulf between the Army and the people, that such an attack was now possible. Like most of the violence that swept the city between night and dawn, the act was gratuitous, the loss of life pointless.

The night was over. So too was the Democracy Movement and everything the students and their supporters hoped to achieve. By taking the Square, the Army had crushed opposition to the Party. It had also shattered the minds of an entire city and an entire generation. The students trudged north to their campuses like people in a nightmare, some hysterical, most in tears, all dazed and exhausted. The dream was over. From now on, it would be the old story: intimidation, settling of scores, silence. All that was left to figure out was how many had died.

6

Counting Bodies
(June 4)

"How many people died last night?"

As the sun rose over Beijing on Sunday, June 4, this was the first question on everyone's lips. Not, "Did the Army really kill people?" Nor even, "Why did they do it?" Simply, "How many?"

It was, and is, a hopeless question. Hopeless because it begged for facts and statistics at a time when no certainties could be established. Hopeless because it refused to imagine the future and wanted only to fix the past. Hopeless too in another sense, for the government did not want the number known and would go to great lengths to prevent the dead from ever being counted. But a number is what people in Beijing needed to know that day. It is what I need to know.

There are no complete statistics, no sound figures. There may never be. There are estimates, and partial hospital body counts, and the endless testimony of eyewitnesses who watched people get shot down, carried bodies into hospitals, saw shirts hung at university gates punctured with lines of bullet holes. There are rumors—of invisible killings, of corpses swept up and burned, of individuals who can no longer be accounted for. Finally, there are the government figures. From none of these quarters comes anything that does justice to the scale of loss or meets the need to know how many really died. Nothing is correct or precise.

Even precise numbers would reveal only one dimension of what the people of Beijing experienced that night. The government's admission that close to three hundred were killed and eight thousand wounded certainly provides numbers good enough to call what happened in Beijing a massacre. But people thought differently on Sunday morning: they were counting the dead in the thousands, not the hundreds. Three thousand quickly emerged as the street statistic, and people clung to it as firm proof that their government viewed them as nothing more than straw dogs to be thrown to the fire.

Do we need to decide between three hundred and three thousand? From a distance, either death toll is atrocious: the number hardly matters. From close up, however, even one death is one too many, and the omission of one in the final count a terrible lie. The quantity of killing matters most to those who died and those who mourn them. Not to be counted is to be lost forever. Numbers may matter less to the rest of us. What matters in a larger sense is the quality of the action that produced the killing, the betrayal of trust that led the Army to open fire in the first place, and to keep on firing long after the first order was given. Any number enumerating the destruction of life is sufficient testimony to that betrayal. The more precise, the better.

Precision has its value. A definite number would help to fix the fact of the Beijing Massacre and bring it out of the limbo of the merely alleged. Otherwise the event seems vague, its shape incomplete and its meaning unexposed. A definite number would stand against the eroding influences of uncertainty and misinformation on which every government relies to keep power in its grasp. In this chapter we will pursue that number through the estimates, body counts, and personal experiences of eyewitnesses. We will search for it, but we will not find it. Nonetheless, the pursuit will instruct. It will tell us something of the nature of the disaster as people experienced it. It will help us trace the contours of the loss.

Where to begin? The experience of eyewitnesses is hardly the place. Many eyewitnesses saw people killed, but many did not. More to the point, none witnessed all the killing. People saw what was around them, at best a few dozen meters in any direction from where they happened to be standing when the guns went off. Their perceptions were muddled by fear and confusion and darkness. How can their individual experiences of the killing be summed into a total?

Nor does the extensive video footage that foreign camera crews shot help to gauge the scale of death. Many viewers thought they saw a massacre on television, but they saw only confused shouting, military vehicles tearing about, a few guns being fired, some bloodied shirts and faces. The slaughter of civilians is not on tape. As BBC correspondent Kate Adie has pointed out, you can't get casualty estimates from video footage. You can't even get casualties. "It happened so quickly and so violently, and the confusion was so great, even the best cameraman in the world could not behave like a feature film cameraman and have the time, the angles and the luxury to record death. It's not like in the movies. Down—one-eighth of a second. People die of shock from a high-velocity bullet at under seven hundred and fifty yards. The cameraman turns—they're gone, down on the floor. You don't see it on the television cameras."

But if journalists couldn't record actual deaths, they could investigate the visual traces of the bloodletting next morning, which is what most did. They found pools of blood at Dongdan, entrails and brain matter on the sidewalks along Changan, and chest-level bullet holes pitting the walls of the Forbidden City. The quality of the actions behind such evidence prompted some on-the-spot observers to assume a volume of killing so enormous that only the largest numbers made any sense. A Dutch journalist was saying six thousand, Soviet journalists were going as high as ten thousand. One of the two eyewitnesses who escaped from the club-wielding soldiers on the plinth of the History Museum and fled into the south city (see the previous chapter) regarded anything lower as unbelievable. "Troops came in from five directions; fought their way all the way to Tiananmen Square; were blocked at every intersection," he points out. "It wouldn't surprise me if we were to find that ten thousand people had died. And that's conservative. I can't be certain that everyone I saw get shot died. But the guns the soldiers were using are lethal, especially if no-one is there to run you off to a hospital right away. When I add up what I saw and all the stories I heard, there's no way the death toll could have been less than ten thousand." The scale of outrage and betrayal that he and so many others felt that night demanded a statistic of equal power. They themselves had seen too many die within their own dozen-meter circles of sight. Many had died, and "many" could not have been dozens, or even hundreds. It had to be more.

The first students from the Square arriving back at their campuses Sunday morning had the same hunch. They had no numbers, only horrendous personal experiences that left them in shock. "Some were hysterical and couldn't stop crying," recalls a Beida campus resident who saw the first contingent return at 6:00A.M. "Some were absolutely confused about what had happened, crying and laughing at the same time. Some just sat catatonic in the trucks, not moving or seeing anything. Others were wild with anger." They brought with them bits of evidence that spoke of something far larger than what lay in their hands. These emblems and experiences they shared these with other students and local residents who came for news of what had happened.

Among the evidence of lethal Army actions were bullet casings that students passed around. "The metal was turned back, like a flower," according to the same campus resident. "People said it was a kind of bullet that explodes when it touches you. Someone told me he saw one of his friends fall to his knees on the Square. He went over to him and noticed a little round red spot on his stomach. He thought, shit, he's been wounded. When he put his arms around him to help him up, he found he'd put his hand right into a big hole in his friend's back. He froze for a few seconds, then had to get up and run and leave the body there."

People saw these terrible wounds and thought that the PLA was using explosive or dumdum bullets, forbidden by the Geneva Convention. In fact, the wounds are in keeping with the conventional type of bullet used in the assault rifles of both PLA and NATO forces. The steel core of this bullet is surrounded by a copper jacket that prevents wear on the steel barrel of the gun. On impact, the bullet sheds the copper jacket, which peels off in a flowerlike burst. The bullet leaves a small penetration where it enters the body, but tears away the flesh where it exits. The wound requires intensive medical treatment. The aim of wounding in this fashion is to impose heavy medical burdens on the enemy in addition to reducing the number of soldiers in action.

At Beida and other campuses as well, students displayed not only bullet casings but corpses. Trucks showed up at Beida at 10:00 A.M. carrying several dead, including a nine-year-old boy whose body was riddled with seven or eight bullets. Where bodies were not available, photographs took their place. The bulletin boards at the front of People's University by noon Sunday were covered with snapshots of the carnage, and bloody shirts and jackets were displayed on tree limbs by the main gate. The same sort of evidence was shown off campus as well. At the Beitaipingzhuang intersection on the north stretch of the 3rd Ring Road, students held up captured weapons and bloodstained clothes to attest to the Army's atrocities.

But the physical evidence was still not enough. People wanted something more abstract, less tied to the particulars of one death that would capture the scale of death and injury: they wanted numbers. By midmorning, estimates of between one and ten thousand dead were in circulation. Most estimates fell between two and three thousand. It felt right to those who had experienced the events of the night before. The range gained currency. An American teacher on the Beishida campus heard three thousand from his students returning from the Square. A Chinese psychologist heard the same number from student orators addressing crowds at an intersection near Beishida. A Canadian journalist heard it from some people standing on the steps of the Beijing Hotel at 8:00 A.M.

Where did the number come from? One of the sources was the Chinese media. The 8:00 A.M. television news broadcast announced that two thousand people had been killed. Radio Beijing's English-language service at 1:00 P.M. said "thousands." The government suppressed these estimates and substituted a much lower number, in the hundreds, but many accepted two thousand as an absolute minimum, and assumed that the total was well over that. A Beida student handbill printed on Sunday morning repeated the television report by declaring that at least two thousand people had died, adding that student deaths were in the hundreds and that the injured ran to the tens of thousands. An announcement over the Beida

Autonomous Students' Union loudspeaker midmorning gave the figure of twenty-one hundred dead. Another Beida handbill printed later Sunday morning accepted three thousand fatalities as the best estimate. A Students' Federation handbill claimed in addition that seven thousand were wounded.

The other source for three thousand—and probably the source on which the television news relied—was the Chinese Red Cross. Early Sunday morning, the Red Cross told foreign news services, including the Canadian Broadcasting Corporation, that twenty-six hundred had died. It based this estimate on reports coming in from hospitals. The Red Cross figure was broadcast Sunday morning over the student federation loudspeakers at People's University. It was also picked up by the Japanese Red Cross. The Chinese government denied it. Even so, an American who went to donate blood at a Red Cross station on the 3rd Ring Road early that afternoon was told that the Red Cross still estimated the number of dead at "a few thousand."

(The American and a British friend had gone to the Red Cross station in response to an emergency appeal for blood that he heard broadcast from an ambulance late Sunday morning. The staff gave them a standing ovation when they arrived, but before they could take their blood, the head nurse informed the foreigners that the Red Cross did not need blood and that they should leave. Was her decision merely an administrative precaution against the fear of AIDS infection from foreigners, or was the Massacre so much an internal affair that China could not bend to accept their donation?)

The Red Cross figure was corroborated Sunday morning by the Swiss ambassador, whose diplomatic duties include representing the International Red Cross. The ambassador went around to hospitals that morning and came up with his own estimate of twenty-seven hundred. Since protocol demands deference to the government to which an ambassador is posted, he was required subsequently to disavow that figure. Nor can the International Red Cross accept it as a matter of public record, because documentation is lacking.

The figure twenty-six hundred cannot be verified. The only hope of reconstructing casualties is to go back to the hospitals on which the Red Cross figure was based and work up from there. Hospitals were collecting and giving out such data on Sunday, but not for long. Doctors on duty Saturday night and Sunday morning would be in a good position to furnish estimates, but few have since dared to relate their personal experiences that night. The subsequent turnover in staff at Beijing hospitals in the year following the Massacre has only further impeded tracking down who was in the emergency rooms of city hospitals when the dead and injured were brought in.

After the Massacre, rumors around Beijing had it that the hospitals were not taken completely by surprise, and that some had been notified to prepare for casualties. A nurse told a Beida student on June 1 that her hospital was ordered to empty all the beds except for patients needing intensive care. Another source in Beijing told me that hospitals were alerted on the afternoon of June 3 to prepare for casualties. Someone whose wife works at a hospital told me that her superior warned her not to go out on the streets that night. I have also heard that hospitals were instructed to take in only cadres and soldiers, not civilians, or to treat soldiers before civilians, and that they were forbidden from filling out death certificates for civilians. The instructions, if given, went unheeded.

In any case, the rumors that hospitals knew what to expect go against the evidence that night. Civilian hospitals may have been forewarned that their services would be needed, but none of those for which I have eyewitness testimony knew the scale of casualties they would receive. Nor did they have sufficient medical supplies, especially blood for transfusions, to handle such an emergency.

Accounts I have from two doctors working in hospitals that night contradict the idea that hospitals were prepared for what was to come. One doctor wrote and published in Hong Kong an anonymous account of his experiences that night. Another agreed to be interviewed only on condition of complete confidentiality. What they describe is almost unimaginable to anyone who has not served as a medical officer in a war. It is with their experiences, rather than their statistics, that I would like to begin.

The first doctor went home from his hospital early Saturday evening. He is careful not to reveal its identity, though it appears to be one of half a dozen in the southern half of the city within three kilometers of the Square. Sometime after 11:00 P.M., he heard the roar of APCs driving at high speed toward Changan Boulevard. The sound of gunfire in his neighborhood shortly after midnight prompted him to return to the hospital. When he arrived he discovered that the first casualties had already been brought in: four or five young people with bullet wounds, plus others who had been struck on the head. The bullet wounds tended to be in muscles rather than internal organs. He set to work removing bullet casings and closing wounds. Most of the wounded were local residents of all ages rather than students. They kept arriving in a steady stream as gunfire in the hospital's vicinity increased.

At 3:00 A.M. "some people rushed in with a man on a stretcher. A woman was holding onto the side of the stretcher and crying. We examined him and found that both his respiration and pulse had already stopped. His pupils were completely dilated. We tried to revive him by means of heart injection and massage, but we couldn't. The three bullet wounds in his chest and abdomen were still bleeding, but he was dead. His wife kept calling his name, grasping at the slim hope that he was still alive."

Two students were brought in at almost the same time with the same wounds. "Their vena cava arteries had been severed and the necks of their thighbones broken. The students who brought them in had tied their shirts and trousers as tourniquets on their upper thighs to slow the bleeding, but blood was still pouring forth. They had all the symptoms of shock: extreme thirst, white faces, and sudden drop in blood pressure. Since we had almost a full staff by this time, we sent them immediately to the operating room."

As the night progressed, the doctor was struck that the size of the copper bullet casings was growing larger. Around midnight they measured 1.5 cm diameter; later they were up to roughly 3 cm. "The wounds were getting proportionately larger and the loss of blood greater. We used up large amounts of gauze and bandage brought down from the other departments in the hospital. What was worse was that our blood supply was running out. Our patients were in shock, and all we could do was use plasma to keep them alive."

Some of the wounded were carried in on stretchers. Others came in ambulances, usually several people at a time. "In one ambulance, five bodies were stacked together. When we took them out we found that three were already dead. Their school badges identifed them as students of Nankai University [in Tianjin], Beijing University, and the University of Science and Technology. The students who came with them said they had been shot by automatic rifle fire in Tiananmen Square. We did our utmost to save the lives of the other two. They had received multiple gunshot wounds in the chest and abdomen." Another overloaded ambulance arrived a short while later, this time carrying seven bodies. "Four of them were already dead. Their bodies were covered in blood. All we could do was close their eyes. By this time the emergency room was in a great mess: blood everywhere, everyone crying and cursing as though we had all gone mad. All the beds were now full, so the newly admitted had to be treated lying on the ground." Only after 5:30, when the shooting diminished, did the flow of wounded to the hospital subside. The staff had handled over a hundred people, of whom sixteen had died.

The second doctor tells a similar story for the Capital Hospital, one of the oldest and largest comprehensive hospitals in Beijing (see Map 3). It was built with money from the Rockefeller Foundation under the name of the Beijing Union Medical College Hospital. After the revolution, it was renamed the Capital Hospital, and though the old name was restored in the 1980s, people in Beijing still call it the Capital. Located two blocks northeast of the Beijing Hotel, it was one of the three hospitals that received the largest numbers of dead and wounded from the Square. (The other two hospitals, Tongren and Beijing, are both to the east of the Square. Beijing Hospital, normally reserved for high officials, began to receive casualties at 2:16 A.M.)

The doctor we interviewed was not on staff at the Capital. He just happened to be in its vicinity about midnight and decided to go there and volunteer in the emergency ward. The first casualties were being brought

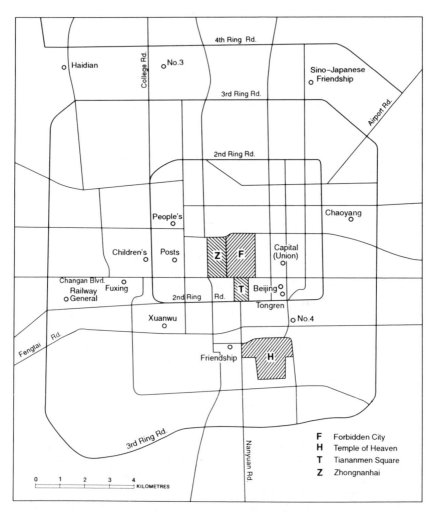

Map 3. Major Beijing Hospitals.

in before 2:00 A.M. Only after the soldiers formed the cordon at the
northeast corner of the Square and started firing at 2:40 A.M. did the trickle
of wounded turn into a flood. Some arrived on the backs of three-wheeled
flatbed bicycles; some were brought in by a bus that changed hands twice
as one driver after another was shot; a few came by ambulance, though
ambulances were not running regularly until after dawn.

"The busiest time was around 5:00 a.m.," the doctor recalls, "just as it
was getting light."

> We heard a lot of gunfire, and not ten minutes later a huge number of
> wounded were brought in. Whereas four had been carried on one flatbed

before, now eight were being brought in, piled double. There was blood all over the place. The guys wheeling in the flatbeds were all doing it on their own initiative. One man I talked to said he had gone back and forth to the hospital six times.

All were suffering from bullet wounds. I saw an X-ray showing a bullet that had gone into someone's shoulder bone. The worst case I saw was a man whose chest had been blown apart. A single bullet had hit him in the back. We rushed him into surgery, but he was dead. We looked for some identification in his bag, but all we found was a notebook with *PLA Daily* written on the cover, plus meal tickets from the *PLA Daily* cafeteria. We decided he was a reporter with the Army newspaper. They sent him down to the morgue. He was the only person I personally saw who was dead on arrival. All the others I received came in wounded, bullet wounds in every case.

I personally handled forty-eight patients, though how many came to the hospital that morning I don't know. The hospital got filled to the point of not being able to take any more. Surgery was really busy. They just couldn't keep up.

The doctor left the Capital about 8:00 A.M., exhausted. Although the hospital staff had seen the worst, the Capital continued to receive gunshot victims through the day. Most of these were brought in from the northeast corner of the Square, where soldiers and citizens faced off in a test of wills that lasted all Sunday. Soldiers had formed a double cordon across every entrance to the Square, but hundreds of people gathered at these approaches, particularly at the northeast corner. Partly it was to see what was going on in the Square, partly to express mute defiance of military control. The standoff between the soldiers' cordon and the citizens' vigil produced a fluctuating stream of casualties. Eyewitnesses watched as soldiers shot down nine people at 9:46 A.M. Twenty more went down at 10:09. Two ambulances appeared almost immediately but were prevented from fetching bodies. The soldiers ambushed one and set it on fire at 10:20. Over the next half hour, two men who tried to drag the wounded to safety were gunned down. This cat-and-mouse game went on throughout the day. By dusk, roughly a hundred people, both civilians and medical personnel, had been shot within sight of the Beijing Hotel. Many of the wounded ended up at the Capital.

Most of the hospitals in Beijing received similarly large numbers of casualties but had neither the beds nor medical supplies to care for them. British journalist Jasper Becker visited People's Hospital, a smaller hospital two kilometers north of Xidan, in the early hours of June 4. "It looked like an abattoir. There were bodies on benches and beds or on blood-soaked mattresses on the floor. Many had gaping bullet wounds on the chest, legs, or head. Students had rescued badly beaten soldiers and we saw one covered in blood who was clearly not going to live." A Beida student encountered much the same scene at 5:15 A.M. when he went to the Posts and Telecommunications Hospital, a facility of 330 beds, a quarter of the distance People's is from Changan Boulevard. Bodies ripped apart by bullets lay in the hallways. The faces of some were beaten beyond recognition. Anesthetics had run out. The staff was exhausted. Only No. 3 Hospital, a large facility in north Beijing, reported a different type of wound: there, students were brought in extensively burned. Otherwise, the story of a high number of gunshot casualties being handled with insufficient resources was universal in the hospitals throughout the city. The only variable was timing: that depended on a hospital's location relative to Army movements.

The first hospital to receive casualties Saturday night had been Fuxing Hospital. A smaller hospital of over a hundred beds, Fuxing is located on the west side of the city near the fateful intersection at Muxidi where shooting first started about 10:30 P.M. A foreign press crew following an ambulance arrived at Fuxing at 11:45 P.M. One member of the crew was later debriefed by Amnesty International. "About every minute, one injured person was brought to the hospital, carried on bicycles or pedicabs. Most of the injured were young people who had been manning the barricades at Muxidi." About an hour after midnight, students counted nineteen dead in the hospital. By 2:45 A.M., the death toll had risen to twenty-six. The student who reported this number also noted that soldiers fired on an ambulance trying to rescue the wounded, and that injured medical personnel were among those treated at Fuxing. At 3:26 A.M., someone called Reuters News Agency from the hospital to say that Fuxing had twenty-eight dead and two hundred wounded. But fighting in that area was not yet over. By 5:00 A.M., according to a student source, the number of dead had risen to fifty-nine. A nurse at the hospital told a Beida student later that Fuxing was filled to the point of being unable to take in any more. This was no exaggeration. A phone call the next morning testified that the hospital parking lot was filled with dead and wounded, the youngest three years old, the oldest in his sixties.

By morning, all Beijing hospitals were besieged by people looking for relatives and friends who had gone missing and were feared injured or dead. Administrators at Xuanwu Hospital, a smaller facility in southwest Beijing, drew up and posted at noon Sunday a list of the dead they had been able to identify. The list gave names and ages. Over fifty were on the list. The eldest was fifty-four years old, the youngest seventeen. The list did not indicate how many corpses the hospital had received but could not identify, nor did it name or number the wounded.

Many people tramped around to the hospitals Sunday morning not to find missing relatives but to count the casualties. People simply wanted to know, and medical staff were sometimes willing to share what they knew, often against the objections of hospital authorities and the police. A history student from Beida was apprehended by police and detained for about ten days for collecting casualty statistics from hospitals. When a Canadian Television crew arrived at the Capital Hospital mid-morning, they found the staff there divided over whether to admit them. A scuffle broke out among the hospital staff, and the doctors and nurses who wanted to let the journalists in ended up having physically to restrain those who wanted to deny them entry. The doctors led the journalists to the hospital morgue to see the thirty or forty bodies there, but the man who had jurisdiction over the morgue refused to open it. Doctors slammed him against the wall, threatening to break the door down if he didn't fetch the key. He agreed to get the key, but instead went to get the hospital authorities, who came down and made sure the television crew left. Before the journalists were ejected, they visited one ward. The patients they saw were all suffering from bullet wounds, mostly in the arms and legs, some in the chest, one in the throat. The journalists were unable to verify the number in the morgue. That evening, though, a nurse at the Capital repeated the number forty.

Table 6–1 shows estimates of the numbers of dead and wounded that I have been able to track down for twelve civilian hospitals and one military hospital on Sunday. The reports were taken at different times that day, some by doctors, some by students, some by foreign journalists. Grossly incomplete, these are less statistics than suggestions regarding the scale of casualties in Beijing. What they suggest is that eleven Beijing hospitals received at least 478 dead on June 4, and that eight hospitals treated over 920 wounded. The first figure fits with what a Beijing doctor found when

TABLE 6–1. Incomplete Casualty Statistics from Selected Beijing Hospitals

Hospital	Location	Number of Beds[1]	Dead	Wounded
Capital[2]	Central	—	40	—
Chaoyang[3]	Northeast	—	—	100*
Children's[4]	West	—	55	100
Erlonglu[5]	West	256	—	40
Friendship[6]	South	—	43	—
Fuxing[7]	West	—	59	200
No. Three[8]	North	—	95	125
People's[9]	West	—	4	105
Posts and Telecommunications[10]	West	330	28	150
Railway General[11]	West	711	85	—
Second Artillery[5]	West	—	4	—
Xuanwu[12]	South	—	50*	—
unnamed[13]	South	—	16	100

Sources: [1]*Quanguo yiyuan minglu*, pp. 1–2. [2]CDP-1022; CDP-0929. [3]A105. [4]ASA 17/60/89, p. 17; CDP-1022. [5]CDP-0622. [6]A109. [7]ASA 17/60/89, p. 17; CDP-0716. [8]CDP-0717. [9]Jasper Becker, London *Guardian*, June 5, 1989. [10]Duke, *The Iron House*, p. 120; A100. [11]ASA 17/60/89, p. 17. [12]CDP-1090. [13]Yige Yisheng, "Yige yisheng yanli," p. 74.

—No figure available.

*Stated by source as incomplete.

he contacted eleven hospitals Sunday morning; he calculated a death toll in those hospitals of over 500.

How many casualties ended up in hospitals other than those listed in Table 6–1? Many, it seems. Eyewitnesses stressed that hospitals all over Beijing on Sunday "were full of wounded and dying people. Even Haidian Hospital, fifteen kilometers from the Square, was full." The volume of casualties was so great that some hospitals had to turn people away. A doctor at People's Hospital told Jasper Becker in the early hours of June 4 that over three hundred had come to his hospital that night but that "most were so bad we sent them on elsewhere." People's was able to handle only a third of those who showed up.

Not all of the 124 medical facilities listed in the 1989 city telephone directory received casualties, nor in the same numbers that those in Table 6–1 did. But many did receive large numbers of dead and wounded. We can generate an estimate by using as a base figure the thirty-two hospitals in the city that ministered to the hunger strikers in May (see p. 37). Suppose these thirty-two handled casualties on Saturday night, and suppose that on average they handled them at the same rate as those for which we have estimates. That would give figures of fourteen hundred dead and thirty-seven hundred wounded by dawn Sunday. Add to this base almost a hundred smaller facilities and clinics, as well as the many military hospitals in Beijing, and the totals could rise to twice that.

Even if we could obtain complete statistics for all these hospitals, we would still not know the total number of casualties. As I noted in the previous chapter, many of the wounded preferred not to show up in hospitals for fear of being charged with counterrevolution. Those with nonfatal wounds preferred to take the medical risk of nursing those wounds out of sight of the public health system over the political risk of exposing themselves to persecution. Only when a wound required surgery or stitching did the risks go the other way. Reluctance to reveal an injury might help explain the imbalance between dead and wounded in the hospital figures. The usual ratio of wounded to dead in gun battles is about ten to one. In Table 6–1, the number of wounded is unexpectedly low.

A corpse could be just as incriminating for his living relatives. The family of a "counterrevolutionary rebel" faced political sanctions and hefty fines. As much as possible, therefore, bodies were disposed of privately or taken out to the countryside for burial. Some delayed removal, however, preferring to keep them for a time as tangible proof of the loss they had suffered and as evidence of the government's cruelty. As late as Tuesday, a man was carting the body of his four-year-old son around the city, lamenting to crowds how a soldier had fired three shots into the child's body.

"This is one of those that Li Peng calls rioters," he wailed. "What crime is this rioter guilty of? Why open fire on someone like this? Is this the people's army?"

Doctors did their best to protect those who came into the hospitals.

The staff in some hospitals simply attended to the injured without keeping a record of who they were. One foreigner heard from a doctor at the Capital that the staff there "bandaged students up as quickly as they could and got them out of the hospital for fear that the soldiers would come in and drag them out." On Sunday morning, in fact, several Capital doctors took two vanloads of injured students to No. 3 Hospital to get them at some greater remove from the Square. It might lower their chances of being implicated in political crimes.

An official publication has subsequently revealed that the Capital also transferred a "student" to the Sino-Japanese Friendship Hospital in northeast Beijing. This young man was in fact a PLA captain being removed for his own safety. He was one of at least thirty-four injured soldiers who were kept secretly in a lab at the Capital until they could be moved to military hospitals. The doctor already quoted who was working that night at the Capital knew of at least nine other soldiers who were admitted, although he personally did not attend to them. While he was on duty, someone came downstairs with a PLA jacket in his hand to announce that the nine had fled, out of uniform. Although many soldiers were brought into civilian hospitals for first treatment, they feared for their safety, and the Army has published accounts of soldiers in civilian hospitals being harassed by hooligans. The staff isolated them in wards out of public view to prevent hostilities from erupting. As soon as possible, they were surreptitiously transferred to military hospitals.

If statistics regarding civilian casualties are hard to come by, information about military casualties is even more hidden. We have only fragmentary evidence of how wounded soldiers were handled. The Army set up emergency medical centers in several locations. The largest was located inside the north entrance of the Great Hall of the People. A smaller center was set up in the Temple of Heaven park in the southeast section of the city, but it had run out of food and medical supplies by Tuesday. The Army sent in medical teams with some of the military columns on Saturday night, but the number of personnel was grossly inadequate for the scale of casualties the Army suffered, and inflicted. We know, for instance, that only seven nurses accompanied the troops as far as the Square that night; others were dispatched but remained in locations further out. The insufficiencies of medical personnel and supplies show either that the Army failed to anticipate the human damage that would occur that night or that it failed to control casualties, including the important matter of where they occurred. Either reading does not redound to the Army's credit. Casualties could have been avoided, and medical support should have been close at hand.

As much as possible, the Army tried to get its casualties to PLA hospitals, of which there are many in and outside the city. Some hospitals, like the one attached to the Second Artillery, China's nuclear force, received civilian casualties, but not in large numbers. The PLA medical establishment in Beijing seemed to have no clear sense of what it should be doing. The Beijing Garrison Hospital near the northeast corner of the 3rd Ring Road received its first military casualty only at dawn when a jeep was able to break out of the blockade at Sanyuanqiao and bring in a wounded soldier. There were few casualties in that part of the city, but all morning it received telephone calls for assistance from other hospitals whose resources were strained. Yet it was not until 11:00 A.M. that the hospital began to organize ambulance relays to fetch wounded soldiers from other hospitals, and only then on its own initiative. By the end of the day it had collected fifty-four soldiers.

Wounded soldiers who could not reach PLA facilities in the first instance came under the medical scrutiny of civilian doctors. By widespread report, soldiers at the Capital Hospital who were given blood tests were found to have amphetamines in their bloodstreams. I have heard this story from three sources. The soldiers reportedly told the doctors that they had been vaccinated to protect them against diseases on the Square, not injected with amphetamines. Some had indeed been vaccinated; it is Chinese Army practice to vaccinate recruits only when they are being moved into areas that require it, not when they are inducted into the Army. But in any case, it is common practice both in Asia and the West to issue amphetamines to combat troops who have been in state of readiness for many days and may be short on sleep. Close-up photographs of PLA troops Saturday night betray symptoms of amphetamine use: extreme tension, darting eyes, dilated pupils, and swollen faces. A vivid example is the bodyguard of Colonel Ji Xingguo, one of the officers who negotiated the students' withdrawal from the Square with Hou Dejian. Ji and his bodyguard were photographed during those negotiations, and the bodyguard's physical features suggest amphetamine use. But this evidence is only circumstantial. Like so much else connected with the Massacre, the question of whether the troops were acting under the influence of amphetamines during the assault has to be left unanswered. The soldiers' state of high tension, which many eyewitnesses remarked on, could simply have been the product of anxiety.

Pursuing numbers that are not available is a fruitless task. Our best estimates are wild guesses. Conclusive evidence eludes us. This is precisely what the Chinese government intends. Indeed, it is instructive not only to seek after the death toll but to ask why an accurate assessment of casualties is not possible. Simply put, the Army and the government have interfered with the evidence from which to determine an accurate assessment of who died. Many means were used.

One way of obscuring the extent of killing was to collect bodies and make them disappear before they could be counted. Some corpses left in the open were collected and removed by truck. On Sunday at 11:20 A.M., a foreign diplomat watched troops in the Jianguomen area load bodies onto trucks; other soldiers kept the crowd away from the bodies to prevent them from being observed. Also on Sunday morning, at least one truck with Capital Iron and Steel markings carrying men in yellow hard hats was sighted on the western side of the 3rd Ring Road removing human remains and taking them westward. Because the rumor on the street was that Capital steelworkers had gotten into a pitched battle with soldiers coming in from the west the night before, what looked to be factory militia may well have been police from the special unit masquerading once again as workers. Sunday and Monday nights, the work of removing new corpses from public places was done by unmarked white morgue trucks that prowled the streets after dark. These trucks did not deliver their contents to civilian hospitals.

Another way of making bodies disappear is to cremate them on the spot, though whether the PLA did this is not known. At 5:58 A.M., almost as soon as the last students were leaving Tiananmen, a column of smoke began to rise from the Square. Another fire started at 6:40. A Beida student watching the Square from a building to the south saw the flames rise and fall as new debris was tossed on. The smoke was thick and black, suggesting that gasoline was used to soak whatever was being burned. It thickened three times between 6:45 and 7:30 A.M. for about five minutes each time.

To those who saw the fires, including a foreign journalist watching from the roof of the Beijing Hotel, it seemed that bodies were being burned. "The smell was overpowering," she remembered. Independently, many intuited that corpses were being cremated. The intuition may have been rooted in people's feelings about what had happened the night before, not in what lay before their eyes. They could not see the bodies burning, but the context of the smoke they did see encouraged them to imagine that corpses were being thrown in the fire. After all, everyone assumed at the time that the carnage on the Square had been of substantial proportions. The disposal of bodies would have to fit in with that assumption: something equally immense, equally disrespectful and degrading. A Qinghua University student perched in a tree at the south end of the Square has stated that at about 6:45 A.M. he could see soldiers heaping plastic body bags together and covering them with a tarpaulin—but he does not say they were then burned. A Beida student leaflet, on the other hand, reported that "the Army used bulldozers to shovel bodies into piles and burn them."

Because the number of people who died on the Square was limited—and many of them were removed by ambulance—the image of corpses bulldozed and burned is far-fetched. Most of the evidence of immediate cremation is circumstantial; plus there are some practical reasons that incineration did not occur. A human body does not burn quickly. Incinerating a body takes several hours at high heat. A fairly reliable Chinese source with access to military information has confirmed that in at least one location—though he doesn't know where—the remains of people who had been crushed by tanks were scraped up and burned with flame-throwers. The unknown location may have been the bridge at Muxidi, where the concrete has burn marks to a depth of about 10 cm and the concrete tiles on the sidewalk to a depth of about half that. Strong burn marks on the concrete paving stones on the Square behind the Martyrs' Monument could also have been made as a result of burning pulped bodies. Other burn marks on the Square, however, suggest the lower heat needed to burn tents and other flammable objects, not bodies. The Army has insisted that nothing other than the debris that had accumulated in the Square was burned that morning, though it has not released the pictures of the fires that military photographers took at the time. The best I can do, once again, is to conclude on a note of ambiguity: the military may have burned some human remains on Sunday morning, but only in small numbers.

There were other reports of flamethrowers having been used, not to burn corpses but to assault living people. One foreign teacher was told Sunday morning by students involved in the previous night's activities that soldiers had cornered citizens in an alley downtown and burned them with flamethrowers. As already noted, students with extensive burns did appear at the No. 3 Hospital in the north end of the city. It would have been a sad irony had they been used, for Chinese technicians developed the first flamethrower at least ten centuries ago. The "fire lance," as it was called, spurted burning gunpowder at the enemy, not to incinerate him but to keep him at a distance.

Yet another measure that people suspected the Army used to make sure the actual scale of the killing would not be known was to kill people out of public view. All evidence of such killing is secondhand, but from two separate sources I have heard that students were executed inside the Workers' Culture Palace. The Culture Palace is a recreation park located to the east of the gate of Tiananmen in one of the forecourts of the Forbidden City. The Army used it as one of its bases of operations against the Square. From several firsthand sources, we know that the Culture Palace served as a temporary detention center for people rounded up Sunday morning. It appears to have been under police supervision, although most of the security personnel there were soldiers. Five foreigners were held there at mid-day Sunday: a British sightseer, an American and his wife, an Italian journalist, and a fifteen-year-old Pakistani boy. The first of these eyewitnesses testified to Amnesty International that police tortured them with electric cattle prods during their interrogations. "During this time he could hear screams from a nearby building into which wounded Chinese detainees were being taken. They had head wounds and other injuries, and they were tied with hands behind their backs and attached to a cord around their necks." The eyewitness got a glance inside this building as he was being taken out of the Culture Palace at 2:00 P.M. It held at least eighty people. "It looked like a butchery," he said. Other cases of arbitrary detention and torture in the Workers' Culture Palace have been reported in the Hong Kong press.

It has been alleged that PLA soldiers not only tortured but executed by rifle butt close to two thousand people on Sunday morning. The story is attributed to the grandson of a prominent Army general, who was among the students taken into the Culture Palace but who was spared when police supervisors learned of his identity. A completely different source tells the same story, but says that the soldiers used knives to execute the detainees. Both agree that they were not killed by bullets. Could execution on this scale have occurred? Probably not. Massacre by hand is an arduous task for soldiers and difficult for commanders to sustain over the length of time needed to put two thousand people to death in this way. The account as it stands can't be true. But given the systematic application of torture and mistreatment, it is likely that some detainees did die at the hands of their jailers. Their deaths will never be made known.

There was, finally, one other way to obscure the scale of human damage after it is over. That was to get rid of any sort of paper trail. This was done. Radiologists in one Beijing hospital were told to destroy all X rays they had taken that night of bullet wounds. There would be no evidence left behind.

The government's most straightforward attempt to obscure what happened was to publish casualty figures that would verify its version of how so few citizens and soldiers died. On June 6, State Council spokesman Yuan Mu presented the government's provisional statistics at an afternoon press conference. Yuan announced that close to three hundred people (both soldiers and civilians) had died. Twenty-three of these fatalities were students from Beijing universities. Roughly another hundred, according to a separate report from the Municipal Party Committee, were soldiers and policemen. Yuan Mu said that two thousand civilians were injured, compared with five thousand soldiers. Mayor Chen Xitong in a speech delivered on June 30 revised the number of injured soldiers and policemen upward to over six thousand, and the number killed to several dozen. He increased the number of student fatalities to thirty-six.

These figures are designed to make several points. One is that the Army suffered more than the citizenry: soldiers allowed themselves to be beaten rather than retaliate with lethal force. It was they who made the greater sacrifice and, by doing so, preserved the honor of the PLA. A second point being made is that the Army did not directly suppress the students, as everyone had feared they would be suppressed. Compared with the thousands of students who took part in the demonstrations, thirty-six is a small number. A final point embedded in these statistics is that the military operation, although it did cause casualties, was on a scale appropriate to the level of opposition the Army had to face. The Army had to use force, and hundreds died, but thousands were not slaughtered, nor were their corpses bulldozed and incinerated on the Square.

Are the government's figures believable? Even my simple table of statistics from eleven hospitals shows that they can't be. Is the government mistaken? Clearly not, for a former PLA officer confided to a foreign friend that the officers working in the Army Intelligence Unit wept when the reports of civilian casualties started coming in. Why, then, has the government issued preposterous data, meanwhile doing everything it can to prevent contrary evidence from ever leaking out?

One possibility is that even the government was appalled by the scale of killing and decided to doctor the statistics to hide the slaughter. The likelier explanation is the need to pacify public opinion. Failing to provide a credible death toll might also serve a purpose. Information that was precise and reliable would not do to domestic public opinion what denial and rumor can: inspire fear. Ambiguity is an effective tool for robbing an opposition of useful ground to stand on. If official statistics are obviously inadequate, then everyone will be reminded, should they have forgotten, that final knowledge, and hence final power, rests with the rulers.

To parade that power, particularly in the provinces, the Chinese government has chosen to police its statistic. It regards anyone who passes numbers other than those given by the government as guilty of rumor-mongering. This is an indictable offense in China. Within two days of the Massacre, few in Beijing dared to make public statements about how many people died. A doctor whose foreign friend tried to raise funds among foreign residents would say on Monday only that his hospital "was crowded with the dead and wounded." In the provinces, where rumor put the death toll at ten thousand, it would take a week to reimpose control. Many mocked the government's statistics and used inflated figures to denounce the regime, only to be subsequently charged for spreading false rumors. The statistics had to be policed.

Chinese emigré intellectuals and students I have talked to since the Massacre have come to doubt the Red Cross estimate of 2,600. They wanted to have a figure as grand as the Army's barbarity, but now, with the Movement in defeat and the Party back in control, they doubt their earlier aspirations and feel that the figures, like their hopes, were exaggerated: "wishful thinking," as one said. Most have confessed to me, as a student from Beida did, that "none of my closest friends were shot." Few can even name names. Accordingly, they doubt the total runs into the thousands. The problem here is less their logic than who they are. As intellectuals, they lived largely separate from the rest of society. Most knew no workers. Yet it was the workers who were the fodder for PLA guns, not the students. Most of the young men who died in the students' cause were unknown to them. It is unlikely that we will ever learn their identities.

But it involves a terrible leap to insist, as the Chinese government has done, that they did not die. At least the Beida student whose friends all escaped with their lives was willing to say that "the total number of people who died is certainly far higher than what the government acknowledges. If Mayor Chen Xitong reported that several dozen soldiers were killed, then simple logic argues that many times more unarmed civilians should have died." Though he would go this far, he still had to round out his comments with a note of caution: "But I become more and more conservative in my estimate."

"I think that the casualties were in the hundreds, not the thousands," another confided to me half a year later. "I would say about three hundred. You have to believe the government's statistic. The Chinese government never lies when it states facts. What it tries to do is hide the facts. But when it says that something happened, then it happened. You have to believe it." Only if you're Chinese, I wanted to say to him. The rest of us get by on a healthy cyncism about what governments want us to believe.

In the war of numbers, the government is winning. People recognize that they do not have the evidence to verify the death count they feel is right. Private estimates become subject more and more to doubt. This doubt, combined with the deep-seated conviction among Chinese that the government never lies about the facts but only misrepresents them, has induced many to err on the side of caution and accept hundreds over thousands. Even the international press has bowed to the shift in the decimal point: the dead are conventionally referred to now as numbering in the hundreds, not the thousands.

This may be the greatest atrocity of all: that we now deny that those who were killed have died, that we abandon whatever truth the state does not assert.

I remain as unconvinced by the official three hundred as I am by the extravagant ten thousand. As always, truth lies somewhere between, but precisely where I cannot tell. Currently I can document only uncertainties. Until information less ambiguous comes along, my fragmentary findings lead me to think that the Red Cross figure of twenty-six hundred early Sunday morning comes closer than any other estimate to the number that died. Perhaps we shall learn the true figure one day, but probably not.

In the end, the number changes nothing. The cold facts are that many died, and that we cannot count their number. The point is not to become lost in disputes over how many were killed but to recognize the enormity of the betrayal that all this killing has welted onto the hearts of the survivors. People have not only died but may not be counted. They have been made to disappear from history. These are the permanent losses of Sunday, June 4.

7

Consequences
(June 4–9)

Within an hour of Sunday's dawn, the People's Liberation Army was in complete control of Tiananmen Square. Lines of armed soldiers cut off all access from the outside while work teams removed the debris of the night's action from the terraces around the Martyrs' Monument and swept it into bonfires. The Square and the Monument—symbols of national identity and moral legitimacy—were no longer in the students' hands. Now they were in the Army's.

The students had succeeded in controlling more than symbols during their seven weeks in the Square. They had won the hearts and minds of the people. By the time the students were forced to withdraw, they could jettison the symbolism of the Square without losing legitimacy in the eyes of the people. The symbols were no longer important, now that the Army had taken them back. The magic they one carried was gone. It left with the last of the students who walked out of the Square.

The Army held this broad expanse of paving stones on Sunday morning, but nothing more than that. The flood of protesting citizens through which it had swept on its way to the Square filled in the streets behind the convoys as soon as they passed. The city was still in citizens' hands, the task of restoring "order" still to be accomplished. Yet as the Army carried out that task, the "turmoil" it was sworn to put an end to was nothing compared to the turmoil it introduced over the following week. Military columns careered dangerously along city thoroughfares. Soldiers fired randomly at cyclists and pedestrians in the street. Public transportation did not operate. No police were on duty. Worst of all, no one knew who, if anyone, was in command of either the Army of the government. Then, at dinnertime on Friday, June 9, Deng Xiaoping appeared on the evening news to congratulate his military officers on the success of their operation. The suppression now had official approval. All hope of some other dispensation, some other outcome, disappeared. Deng Xiaoping was in control, Li Peng still in power, and the Beijing Massacre a justifiable act. These were the narrow terms under which the people of Beijing henceforth would have to live.

Applying a military solution to a political problem, as the Communist Party did in this case, leads to consequences that are not always foreseen. My concern in this chapter is to probe some of the consequences of the Army's entry into Beijing that surfaced during the week between the assault and its justification. One was the spread of military control over the entire city. Another was the enormous destruction of military equipment. A third was the students' preparation to face what they assumed had to come, the Army's takeover of the campuses. There was, finally, the most disturbing consequence of all: the threat of armed conflict between PLA units. None of these things had to happen after the Army took Tiananmen Square, but they all did. Between Sunday morning and Friday night, the citizens of Beijing found they had much to fear.

The immediate target of the military assault on the night of Saturday, June 3, was the capture of Tiananmen Square. The PLA realized, however, that controlling the Square necessitated controlling far more than the Square itself. The citizens of Beijing were too hostile toward military intervention to accept their fate lying down. Tiananmen would not be secure unless the Army secured the approaches to it, and the approaches would not be secure unless the Army secured the entire city, so great was the fear of reprisal and counterattack. A mission intended to remove demonstrators from one public space thus had to be, at the same time, an operation designed to remove everyone from all public spaces. A one-night's attack had to broaden into a full military occupation. And so for the next four days, reinforcements poured into Beijing and the Army made its presence felt throughout Beijing. Armed sentries would remain to patrol the city for the next eight months.

Large detachments of personnel and armor began rolling into Beijing on the heels of the assault force before dawn on Sunday. The first major convoy entered from the east, breaking through the makeshift barrier at the 3rd Ring Road at 5:15 A.M. The convoy consisted of eight tanks, twenty-one APCs (including at least one command-post APC), and as many as two hundred troop trucks, supply trucks, jeeps, and ambulances. It set out from a base well east of the city early Saturday evening. It was blocked for much of the night in the far eastern suburbs, but got going again before 5:00 A.M. Soldiers on board fired and threw smoke bombs as the vehicles approached the Jianguomen overpass over the 2nd Ring Road at 5:20, but people on the street still dared to shower the military vehicles with bricks. The attack caused an ambulance toward the end of the column to crash. Two soldiers in the cab attempted to run but were apprehended and beaten; their vehicle was set afire. Cyclists from as far away at the Beijing Hotel rushed toward the column in the hope of stopping it, but by the time the tanks were a block east of the hotel, ten minutes later, everyone in the street panicked and fled from their path.

In a quick farewell gesture of defiance, some demonstrators pushed a bus across Changan just east of the Square and set it on fire to toughen the feeble barricade of two front-loaders already positioned there. It was only a gesture, too insubstantial to resist armored vehicles. Within minutes, five tanks and the command-post APC reached the intersection. Rather than going around the barricade, Tank 407 drove straight through the front-loaders, crunching a path between them.

"The barricades were flimsy that night because the people weren't ready," observed a journalist who watched the armored vehicles from the Beijing Hotel. "Earlier there were 6 heavy trucks filled with cement and coal. On Saturday night they weren't ready. People pulled traffic barriers across the road but the tanks just went right through them. The two little earth-lifters were only symbolic, they didn't cover the whole width of the road. When the column of tanks arrived, the lead tank went right through them. The other tanks didn't do that, they went around them. There was no need for the first tank to destroy them, but it did."

By 5:30 A.M., ten tanks and one jeep had passed through the barricade. Within a minute they were followed by open trucks, from which troops were firing heavily and indiscriminately. At 5:34 about thirty supply trucks, some carrying forty-five-gallon fuel drums (to refuel the tanks), had passed toward the Square. Two minutes later, a truck in the column stalled and a soldier jumped out of the cab to try to get it going again. Bystanders raced toward him, throwing rocks. He retreated to his cab but could not defend himself. Rocks shattered the window and he was stoned to death within a minute.

At the same moment, another group of APCs came south toward the Square along the west side of the Beijing Hotel, AK-47s blazing. A foreigner who saw them coming leapt over the fence into the hotel compound but found the hotel doors locked. "I threw myself underneath a car in the parking lot with some other people. The soldiers were not just shooting at people to drive them off. They were going after anyone in sight. Then the big tanks went by. They also strafed us again and again. Everybody was pinned down. After maybe five minutes of this shooting, the tanks were in the Square. The windows of the cars we'd been under were shattered."

Much heavy armor poured into the city just after dawn. An official account of one tank unit indicates that its deadline to be in position at the north end of the Square was 5:40 A.M. That unit had to plow through a dozen barricades, yet it arrived twenty minutes ahead of schedule. Other units fared less well. Many tanks were attacked and some destroyed. Five of the tanks in the large resupply column were stopped and burned by crowds even before they got to the 3rd Ring Road. Obstacles interfered with their tracks, the crews abandoned them, and people threw Molotov cocktails inside. It was rumored that the soldiers who abandoned their tanks took off their uniforms and threw them into the flames. Significantly, the tanks did not blow up, which means that they were not carrying ammunition in their turrets. Probably none of the tanks coming in from the east at dawn were armed. Their role at this point was pure intimidation. Supply trucks in the convoy may, however, have been carrying tank shells.

Other armored units failed to make it to the Square for another reason: they got lost. A group of fifteen tanks and one armored recovery vehicle (an ARV is used for towing damaged tanks), followed in turn by soldiers on bicycles, came south down the west side of the Beijing Hotel but turned east instead of west, as they should have done to get to the Square. Another group of twenty tanks, eighteen APCs, fifty-three trucks, and an armored fuel tanker neglected to turn at all, crossing Changan and continuing straight south. It took both groups an hour to find the Square, where they drew up in formation at 6:47 at the north end behind the soldiers' cordon. According to a Chinese student watching from a building near the Square, the Army by 7:00 had over a hundred tanks and other armored vehicles in and around the Square. A Japanese reporter counted ninety-six tanks and sixty-five APCs.

Armored convoys also entered Beijing from the west, though different tanks were used. The guns on the tanks coming from west had 100 mm bores, copied directly from a Soviet model. They can be recognized by the position of their fume extractors, the short piece at the end of the gun that is wider than the rest of the barrel. The tanks with fume extractors at the end appear to have been from the 6th Tank Division, stationed northwest of Beijing and linked to the 38th Group Army. Those coming from the east had modified 105 mm guns, on which the fume extractor is located at the middle of the barrel, not at the end. These tanks appear to have belonged to the 1st Tank Division, which had been brought up from Tianjin to the east and whose command is coordinated with the 27th Group Army.

Once inside the city, the tanks seemed to be deployed solely to cow the civilian population, unsuccessfully. At 5:20 A.M. there was still a crowd of more than several hundred protesting in front of Zhongnanhai. Thirty-two tanks driving four abreast were dispatched from the Square to end the protest. They fired tear gas to disperse the demonstrators, then drew up in a defensive formation around the front gate. One of these may have been the tank that crushed and killed eleven people on their exodus out of the Square. The tanks' initimidation value was clearly limited, though, for by 8:00 A.M. an even larger crowd had gathered just west beyond Zhongnanhai at the Liubukou intersection.

As armor came into the city in this final wave of military assault, troops already inside the city were deployed to take control of strategic points around the downtown area. The 2nd Ring Road marked the outer limit of their control on Sunday. Its overpasses were in soldiers' hands by dawn. A Chinese eyewitness traveling the 2nd Ring Road that morning confirmed that the overpasses were under Army control but that citizens continued to cross them with impunity. "You could cross, but you couldn't stop. You had to get across quickly." Military control did not discourage people from blocking reinforcements in the immediate vicinity. A unit sent to replace the guards at the Chaoyangmen overpass on the east stretch of the 2nd Ring Road was stopped by crowds a mere five hundred meters short of the overpass Sunday morning.

The Army's control of the Square was just as circumscribed as its control of the ring-road overpasses. A tight cordon of soldiers with permission to fire was positioned at each roadway leading into the Square before dawn, but the Army's jurisdiction went no farther than the four-hundred-meter range of its AK-47s. Crowds gathered at all these points of entrance around the Square to see what was going on and keep sounding the voice of protest against the military occupation of their city. At the south end of the Square at Qianmen, a group of young men on bicycles cycled within range of the soldiers whenever news of atrocities reached them, yelling insults. At the northwest and northeast corners, a thousand people kept vigil before the hundreds of soldiers lined up against them.

At the northeast corner, the soldiers' intermittent firing that had started at 2:40 A.M. Sunday morning continued until 9:00 P.M. A young man in blood-soaked clothes told an eyewitness late Sunday morning that he had been there since the first shots were fired, ferrying the casualties out. "Don't be afraid," he told her. "This isn't my blood. It's the blood of 173 people. I run to the front every time I hear gunfire to carry away those who get shot. I have helped to carry 173 dead and wounded. A lot of them died in my arms." As he spoke, he shook and cried. "It's useless to go on living in such a terrible world. But I want to rescue as many people as I can before I die."

As soldiers and citizens threatened each other around the perimeter of Tiananmen Square, helicopters buzzed overhead. Their presence in the sky made people feel vulnerable, and word passed around that snipers in helicopters were firing from the air into residential courtyards on the east side of the city Sunday morning. This charge cannot be confirmed and seems unlikely. The real focus of helicopter activity was Tiananmen.

A helicopter landed in the Square at 6:56 A.M., the first of a long relay of helicopters through the day. They were bringing in food supplies.

(Many units came to the Square without food and others lost the food they brought with them in confrontations with citizens. What the helicopters could transport in was so limited that by Monday the soldiers were down to a daily ration of two biscuits.) On their return they carried wounded soldiers to military hospitals outside the city. Rumor had it that civilian corpses were also being flown out. A confidential source at an Army airfield west of the city considers the accusation unlikely, for he saw wounded men removed from the helicopters by stretcher and put into ambulances that sped away with lights flashing. Corpses don't require emergency treatment.

Elsewhere around the city, mobile units of soldiers kept crowds at a distance by firing into the air. Sometimes the gun barrels were aimed lower. No one felt safe when the Army passed. Two foreigners walking northeast of the Beijing Hotel on Dongdan were suddenly overtaken by such a convoy about 7:00 A.M.. "Suddenly shooting started. People in a shop with the door open hustled us in. Forty or more vehicles—troop trucks, tanks, APCs—headed north past us, firing as they drove."

Outside the downtown area, citizens still maintained barricades on Sunday, and not all Army convoys opened fire at first to clear a path. A column of eight troop trucks and about three dozen APCs coming in from the north was stopped at the intersection of College Road and the 4th Ring Road at 11:00 A.M. Sunday. Only after four hours did the soldiers receive the order to open fire. Under a screen of gunfire that lasted almost twenty minutes, the APCs broke out of the blockade. The trucks did not, however. Left behind, their burned hulks could be seen there the next day.

Another protective strategy that PLA units in the city used, besides opening fire, was driving at high speed. An eyewitness at the Beitaipingzhuang intersection on the north stretch of the 3rd Ring Road saw a convoy of a dozen military vehicles speed by at 2:30 P.M. Sunday. The stoplight at Beitaipingzhuang turned red, but the column roared straight through without stopping. People crossing at the intersection scrambled to get out of their way. A girl on a bicycle didn't make it. The fourth or fifth truck in the convoy knocked her down. The column was suddenly surrounded. Furious spectators began throwing things. The commander in the jeep at the end of the column ordered a soldier riding with him to fire into the air to scare the people away. Everyone dashed for cover and the column passed.

Sunday night the Army's presence in the city continued to grow, as giant convoys of troop trucks and armor moved incessantly in and out of the city from both the east and the west. An American woman at the Beijing Hotel heard the sound of tanks first approaching late in the evening. "10:35 P.M." begins the entry in her diary. "Presumably I hear the tanks coming up. I do. I'm deeply frightened. They patrol East Changan until we go to sleep. That night I dream I am descending a metal staircase thousands of feet above the ground. I turn to go down it backwards, ladder-style, so I won't have to look at my destination. It metamorphosizes into a lacy wrought-iron ladder; through its filigree I can see the distant ground and feel vertigo coming on; I will fall."

Even larger columns continued to come into the city as she slept, with soldiers riding shotgun. An observer at the Dongdan intersection watched a convoy of over sixty tanks and APCs enter from the east at 2:00 A.M., followed by another thirty-plus armored vehicles two hours later. More convoys followed during the day on Monday, some of these accompanied by squads of armed foot soldiers deployed along their flanks. Their fire seemed nervous and indiscriminate. The high rises in the diplomatic compounds on East Changan Boulevard beyond Jianguomen made them particularly trigger-happy. A column of troop trucks coming over the Jianguomen overpass on Monday had to drive past burning vehicles churning up thick black smoke. An unnerved foreigner observed it from his apartment. "The soldiers fired into the air as they drove through. As they came out the other side of the billowing black smoke they were still firing, but now the guns were down and they were firing into the alleys on the west side of the ring road."

Apartments on the west side of the diplomatic compound facing the 2nd Ring Road were first hit by gunfire Monday afternoon. A convoy coming down the 2nd Ring Road saw that protesters were torching vehicles about half a kilometer north of the Jianguomen overpass. The soldiers on the trucks fired in the air to scare off potential attackers as they drove through. Because most of the people were on the east side of the ring road, with the diplomatic compound to their backs, the soldiers fired east over the people's heads. Between fifty and a hundred bullets hit the west side of the foreigners' apartments. It was not the soldiers' intention that the diplomatic compound come under fire, and no one was hurt. Still, for foreigners and Chinese alike, splashing bullets on diplomatic apartments was an ominous sign that the Army might no longer be under the supervision of the government—that, in effect, China was no longer being governed.

Foreigners could choose to leave when their homes came under fire. Most Chinese could not. They had no option but to stay put and live with the military's presence. Not surprisingly, street resistance dwindled as the days wore on. On Sunday morning, people were still attacking unprotected soldiers and undercover agents. A Chinese resident was out on his bicycle on the western side of the city just inside the 2nd Ring Road at 7:00 A.M. when he came upon two men lying in the intersection near the White Dagoba Monastery, badly beaten. "The people there claimed they were plainclothes policemen. One was definitely dead, and the second one was moaning. I think he was on the point of dying. They said he'd been lying there for hours, and nobody would come to assist him." By Monday the Army was too powerful, and the people too fearful, to permit gestures of this violence to occur. The last organized resistance on the streets faded

out. All that was left were the individual symbolic acts: burning an abandoned bus, dragging a garbage can into the middle of the road, cursing
work teams dragooned into cleaning up the debris on the streets. Cyclists
passing such a work detail in Fengtai on Monday shouted, "You bastards
call yourself Chinese?"

The most memorable piece of defiance on Monday, broadcast to the
world, was the act of a lone nineteen-year-old who walked out into the
middle of East Changan Boulevard near the Beijing Hotel in broad daylight and stopped a column of seventeen tanks heading east out of the
Square. Carrying a light jacket and a small satchel, he decided on the
spur of the moment to talk reason with the military. When the lead tank
stopped, he climbed on the turret and banged on the hatch. The tank commander eventually emerged to speak with him. After he jumped down, the
tank gunned its engines to move forward, but he skipped back into its path
and forced it to halt a second time. At this point, half a dozen young men
ran out from the crowd, waving to the tank to pause long enough for them
to get this young man out of harm's way. Wang Weilin, son of a factory
worker, was reportedly sentenced to ten years' imprisonment for his nonviolent intervention.

Thenceforth, the people of Beijing could do little. Although students
stopped a column of Army trucks on Monday, few dared repeat the tactics
of the previous week against a soldiery armed and authorized to shoot.
Young men were still manning barricades at every intersection on the
northern section 3rd Ring Road, except the overpass over by Airport
Road, where large numbers of soldiers were bivouacked, but the citizens
would lose control of the 3rd Ring Road to the Army Monday night. The
last resistance was collapsing; it was impossible to continue. On Tuesday,
there was talk of calling a nationwide strike for noon Wednesday, but it
was unreasonable to think that workers would dare rise against the new
forces arrayed around them.

The Army did not let up its random intimidation of the city's residents
even after all resistance had crumbled. Early Wednesday morning, sometime before 4:00 A.M., a foreigner overlooking Changan Boulevard on the
east side of the city watched several hundred armed soldiers search the
diplomatic quarter with flashlights.

> They scouted the Jianguomenwai area, catching people and dragging them
> out in front of the Friendship Store. Some poor bastard came over the
> overpass on his bike. Six or seven soldiers with AK-47s raced up to him,
> yelling like dogs barking. They scared him so much he almost fell off his
> bicycle. He threw his hands up in the air, but they knocked him to the
> ground and beat him with their rifles. I heard him yelling, "I have a wife!"
> They dragged him screaming toward the bushes on the south side of the
> road, threw him down and beat him again with their rifles, then dragged him
> over in front of the Friendship Store.
>
> After about a half an hour of searching, the soldiers had collected a dozen

people. They made them kneel in a row across Changan Boulevard facing west. Soldiers behind them put guns to the backs of their heads. They put a jeep in front of the them, and the driver crept along, bringing the headlights across the group while a photographer snapped pictures. They did this twice. The ordeal on their knees lasted three-quarters of an hour before they were loaded into trucks and taken away.

The onslaught against the little society on the Square had become an onslaught against all Beijing. The people were the enemy.

The Army did suffer a major loss in the first days after the Massacre—other than its respect and credibility. It lost equipment, in tremendous amounts. In practically every location around the city, military vehicles were abandoned and set on fire. They burned in the hundreds through Sunday and Monday, a surprising consequence of the Army's entry into the city.

In those first few days, the Army could not extend its control over much more than Tiananmen Square and some of the 2nd Ring Road overpasses. That left smaller convoys, especially if they stopped in the street, vulnerable to incendiary attack. The longest line of burning vehicles on Sunday was in the area toward the western end of Changan Boulevard at Muxidi, where the fighting had been so fierce the night before. From there to Liubukou, as a Beida student described it for me later, "there were burned-out armored vehicles, trucks, and buses. The boulevard was entirely in the control of the Beijing people. Only the area right around Tiananmen Square was in the hands of the soldiers. Everybody was out in the street discussing what had happened the night before and damning the government."

Still unburned at Muxidi was a long line of thirty-seven APCs, part of a larger convoy of military vehicles that had been stopped by a bus barricade on Changan Boulevard. The convoy also included eighteen trucks, eight armored trucks, two ARVs, and about twenty jeeps and other small vehicles. Shortly after 7:00 A.M., the lead APC had crashed into the bus, and the second APC had plowed into the first. A foreign eyewitness arrived on the scene about two hours later to find the APCs sitting hatch-down and motionless on the street. "The bus was already burnt out, and the first APC was burning. Alongside the banks of the canal and around the burning vehicles there were thousands of people. The street and the sidewalks were covered with stones people had picked out from the sidewalks to throw at the tanks during the previous night. The street was covered with traces of blood." Some people were standing on the APCs to get a better view.

"As we approached the bridge at Muxidi, a soldier suddenly emerged from under it together with another young person. When he was discovered, the mood flared up. I heard cries of 'Beat him!' and 'Kill him!' People on our side of the canal began to throw stones. On the other side, where the soldier was, a large and threatening crowd gathered as he came up the bank. He was wounded or burned. The man seemed doomed." The eyewitness's Chinese companion turned away.

"I don't want to see this," she whispered.

The worst did not happen, however. The young man accompanying the soldier waved back the crowds and spoke in his defense. A few joined in. "Don't throw stones!" "Don't beat him." The soldier was led away to safety to a compound south of the bridge by the canal.

Attention turned back to the street. Without any warning, the APCs started up their engines. They were not empty, as everyone had assumed, but had been waiting, presumably for orders. One of the APCs tried several times to pull out of its position and head back toward the west, but its path was blocked by too many burning vehicles. Exit was impossible.

> A young man climbed on another APC which was standing next to a burning vehicle and must have been quite heated up. He tapped at one of the top hatches, speaking to the soldiers inside. A few minutes later the hatch was opened and six or seven soldiers emerged. They took off their helmets and handed over their weapons. I heard cries of "Kill them!" and people rushed over. Others shouted "Don't hit them!" The soldiers were guided away to a building on the other side of the street. A few minutes later, the APC next to the first opened the top lid, and a hand appeared waving a white piece of cloth. Soldiers emerged and gave their automatic rifles to some young men. They hugged. From what we heard afterwards, the guns were not loaded. In rapid succession the other APCs opened and more soldiers emerged. They were all quickly guided away from the crowd by a student group into buildings across the street.

According to a PLA account, some of the soldiers were beaten and were admitted to No. 307 Hospital. The Hong Kong newspaper *Ming pao* two days later stated that this column belonged to an army within the Beijing Military Region, and that the regiment commander was ordered executed for refusing to fire on the people.

People in the crowd got inside a few of the APCs and tried to drive them. After some ammunition containers and heavy machine guns were removed, the rest were torched. Ammunition bursts could be heard in some of the lead APCs; those further back in the column seemed not to have ammunition stored inside. By the end of the day, the stretch of road leading up to Muxidi was littered with the blackened carcasses of burned machines. So too were many of the other approach roads into the city.

Mayor Chen Xitong announced in a speech on June 30 that more than 1,280 vehicles had been destroyed: 1,000 Army trucks, over 60 APCs, 30-odd police vans, plus some 120 public buses and 70 other vehicles. His statistics were meant to dramatize the violence and the losses suffered by the PLA. The problem with the argument that unruly mobs destroyed the equipment of restrained soldiers is that it confuses cause and symptom. More than half of those vehicles were destroyed not in the heat of battle but after the Army had taken the Square. The destruction was a consequence of the Army's incompetence after the invasion, not a sign that civil order had collapsed and needed to be restored by armed force.

What amazed many observers was that the vehicle-burning went on unimpeded through Sunday and Monday in broad daylight. Many watched in amazement as young men torched disabled vehicles around the city without fear. A foreigner in the Jianguomenwai diplomatic compound on Monday watched "somebody go out and light up one of the trucks on the 2nd Ring Road. He walked out on the blind side of the truck [away from sentries], then just leaned into the cab, and it lit up. He must have tossed some gasoline in the cab to light it." Another diplomat was surprised that "the Army was still there and just watched their own trucks being burned."

The soldiers appeared to take little interest in arson, and few precautions against it. Sentries were posted by some disabled vehicles when they were close to troop concentrations, but otherwise the Army seemed to write off any vehicle that broke down, except to cannibalize occasionally for parts. Perhaps there were just too many broken down or disabled vehicles to keep track of. One foreigner estimated the PLA truck breakdown rate at one in fifteen.

Was it exceptional bravery on the part of arsonists, or soldiers' indifference that allowed this destruction to go on? Was the PLA really so incapable of occupying the city in a disciplined fashion and keeping the populace at bay for the first thirty-six hours? Or was it something else?

Many charged as early as Sunday morning that the arson was an Army plot to exaggerate the scale of destruction for propaganda purposes. People declared they had seen soldiers destroying their own equipment "to give the government an excuse for cracking down." To a Chinese informant, a Supreme Court employee on Sunday described seeing six armored vehicles destroyed in just this fashion on the west side of the city. The crews stopped the vehicles themselves, got out, poured gasoline over them, moved back, and then set them on fire by shooting them. Similarly, a bystander looking at the burned military vehicles at Muxidi Sunday morning told an eyewitness, "Look, we didn't do that. The soldiers did it." This way of explaining the destruction has appeal when other logic seems to evaporate. It is plausible that the government found it useful to burn vehicles, especially during the daylight hours when better photographs could be taken. Still, an APC is too expensive a piece of equipment to throw away as a photo prop.

Another explanation for the destruction, or at least for the apparent indifference of the soldiers, is that the Army, or certain units, saw the operation as a chance to get rid of old and outdated equipment. This rumor became widespread. Some people insist that they saw soldiers burning

Army vehicles as late as Tuesday. A payoff logic was invoked to explain this. Having come to the defense of the Li Peng government, the Army could reasonably expect that the government would not stint in paying for the costs of this operation; into those costs could be written the price of a thousand new trucks and five dozen APCs. Perhaps this logic credits the Army with too much cunning. It also ignores the fact that not all the equipment people destroyed was outdated.

The simplest explanation is that the most of the destruction was the work of angry citizens who had no other way of fighting back. If soldiers contributed, it was more through carelessness than intention. The destruction, in other words, was simply a consequence of an army's invading a city it could not secure.

While citizens burned abandoned vehicles on Sunday, students set about organizing the defense of their campuses. They expected, as a consequence of the takeover of the Square, that the Army would move next to occupy the universities.

The state of anticipation on the campuses was painfully high Sunday morning. Even before dawn, students there were preparing for an attack, barricading the main gate and placing lookouts with flashlights on buildings to signal the Army's approach to the students below. Some were willing to give up nonviolence and struggle to the death. At Beishida, Molotov cocktails were stockpiled in the building closest to the main gate, to be lobbed from the roof onto whatever assault force the Army sent against them. At People's University, the student broadcasting system announced to the crowd of close to ten thousand gathered at the front gate at 7:00 A.M. that, according to a foreign student who had heard it, "thousands of people had been killed, the Army was coming to the university, and they were only ten minutes away. And they kept broadcasting this ten-minute warning over and over." As if to fulfill the prophecy, two black cars appeared from the north.

"They're from the Army!" someone yelled, and the crowd surrounded them in an instant. "One car got flipped on its side. People were bashing them, then started to light them on fire. The people inside crawled out and amazingly no-one was hurt." The prophesy was not fulfilled; the men who emerged were teachers coming from Beida.

As People's University braced itself for the arrival of troops, word went around that Qinghua University was already occupied by the PLA, and that Beida was surrounded by troops about to move in. The time of reckoning seemed at last to be looming in the early afternoon when an APC approached People's from the south. Oddly enough, though, it stopped in orderly fashion for a red light at the intersection to the south before proceeding toward the main gate.

"There was an instant of panic," recalls a foreign eyewitness. "People scrambled in all directions, fearing it was the vanguard of the PLA force that everyone expected was coming to occupy the university district. It turned out, to the elation of the crowd, to be a captured APC piloted by students. It was draped in large part with red banners, one student standing up in the turret hatch with wind goggles and leather helmet on, another posing at the machine gun with bandoliers of shells across his chest." This was one of the long line of APCs stalled at Muxidi earlier that morning.

Student activists did not wait passively for the Army to turn up. They also took the initiative in what they knew would be a coming propaganda war with the government and produced as many leaflets as their printing machines could turn out before the troops arrived. The students' printing shop at Beida was busy through Sunday morning churning out anything from neatly typeset declarations of purpose to hastily scrawled newssheets and handbills. They carried two basic messages. One was to spread the truth of what happened. "An Appeal to the International Community" pleaded to Chinese both at home and abroad "to use all possible methods to let the people of our country and the whole world know what really happened on this dark and tragic night." The other message was to mount resistance against the regime. "The June Third Atrocity," after reporting briefly on the night's events, called on workers to strike, students to boycott classes, and merchants to suspend business.

A foreigner at Beida went to the students' print shop midmorning looking for a friend who was working there. "It was so busy there, you can't imagine the atmosphere. There were fifty people inside this small room, piles of paper, at least ten Gestetner machines, two or three photocopy machines. Everyone was rushing, rushing, rushing. I tried to convince my friend to leave with me for fear that the soldiers would come and find her there. She said she couldn't leave until the work was finished, but promised to leave when it was done. I couldn't force her." The students feared that the soldiers might come any time and realized that they could not continue printing forever, even for the rest of the day. "Late in the morning I went back, and the place was empty. Nothing. No printing machines, no Gestetners, no photocopy machines, nothing. Empty." All traces had been carefully kicked over.

Some students at Beida employed their energies working out defense strategies for their campuses, digging foxholes around the gates and stockpiling stones and Molotov cocktails. But these defenses, like the print shop, lasted less than a day. On Monday, the students decided it would be useless to defend the campuses against the Army. The only reasonable course now was to make all signs of resistance disappear so as to give the government as little excuse as possible to punish university students. Those who chose to remain on campus did so simply to register a silent protest when the PLA arrived, not to fight back. Their presence would be their sole resistance. Many expected to die. A foreign teacher at Beishida recalls an eerie tutorial he had with a graduating student just after lunch on

Sunday. "I had previously arranged an appointment with this student to polish his senior thesis, and he showed up. I asked why he was still here. He said that he had no family in the city, and had been involved in the demonstrations from the beginning and was not going to run. He said that they would be on the campus at 1:30 p.m. looking for students, but he wanted to make sure he finished the assignment." In his own words, "I want to write a good paper before I die."

The Army did not come to round up student activists at 1:30 P.M. as promised. Students then decided that the Army was waiting for a night attack. The mood that evening "was extremely tense," recalls a student at Qinghua. "Terrifying. Everyone thought that that night or the following morning the Army would come onto the campus. There was great anxiety." Yet nothing happened, other than a few tear-gas canisters that passing soldiers lobbed near the Beishida front gate early Sunday evening to keep the crowd there from attacking an Army vehicle as it passed. Sunday's rumor of an impending attack on Beishida was repeated on Monday; it mentioned 3:00 P.M., but again nothing happened. The Army left the campuses alone. Troops did come onto the campus of the Beijing College of Chinese Medicine in the northeast quadrant of the city between 3:30 and 5:00 A.M. Tueday morning, pursuing demonstrators fleeing from one of the last barricades on the 3rd Ring Road, but university students were left alone.

By Monday evening, there seemed no point in waiting for an attack that would never come. Most students decided that it was time to leave the campuses. Self-sacrifice seemed pointless now that the Movement had been broken. Some found shelter in citizens' homes within Beijing, others slipped out of the city by whatever transportation was running. (A Beishida student from Tianjin who did not leave his campus until Monday afternoon feared detection on the trains and chose to go home, 150 kilometers away, on a bicycle.) Some responded to the call of the Students' Autonomous Federation to quit the campuses and carry the struggle into the provinces. One of these, a Beida student from distant Xinjiang province, tried to smuggle leaflets by train from the Xizhimen Railway Station on Wednesday but was caught and incarcerated for several months.

On Wednesday, university administrators at Beida and elsewhere encouraged the last students to leave the campus. Better to depart than wait to be arrested; better also not to tempt the Army to come onto the campuses with the excuse of disciplining the students. Those who departed left a few small tokens of remembrance. Several cones of sand a few feet high stood in the middle of the main road leading south from People's University. They looked like what they were intended to resemble, a set of traditional graves.

The goal of dispersing the students—if that indeed was a goal—was thus achieved without effort on the Army's part. The Army's only reason for going onto the campuses would have been to prevent students who might have gotten hold of guns from counterattacking. This was not a groundless fear. Weapons had been lost in considerable numbers (only a small proportion of them were returned), and many students know how to handle guns (fifteen hundred Beijing college students had undergone national defense training over the previous two springs). The question for the Army was whether any of the arms the PLA lost during the operation had ended up in student hands. Students at one campus in northwest Beijing (to remain unnamed) told a Canadian reporter Monday night that they had PLA weapons, though he was not permitted to see them. Students on another campus told a foreigner that they had buried captured weapons. On the other hand, a student at yet another campus complained that they had none and naively asked American students whether their embassy might supply them with guns! All these statements may have been sheer bravado. In the end, so far as I know, no student turned a gun against the PLA.

The only incident involving PLA guns on campuses occurred on Sunday afternoon at Beishida. I have heard of the incident from several sources, none of whom agrees with the others about precisely what happened, but the general outlines of the story are clear enough.

In the middle of the afternoon, two Army trucks drove through the side gate of Beishida carrying arms and a small number of soldiers. One eyewitness described the soldiers as looking "very frightened and confused, and young." The soldiers said that they were deserting and wanted to give the students guns to arm themselves. The students deliberated as to whether they should accept the forty or fifty rifles being offered to them. They did. Student representatives then climbed into the PLA trucks and went back with the soldiers to their unit to persuade others to abandon the defense of the regime. One eyewitness recalls seeing them drive north from the campus about 3:00 P.M., soldiers at the wheel and students on the hood holding up banners. Later in the afternoon, the students turned the cache of rifles over to the municipal police. Some said it was because the rifles had no firing pins, others that it was because the Martial Law Command got wind of the incident and threatened the university authorities with an attack if they weren't returned.

The key element in this story is not simply that PLA soldiers attempted to create an alliance with the students. It is that they said they were members of the 40th Group Army, and warned that the 27th Group Army, which they identified as the main culprit in the previous night's violence, was about to attack the campus. This incident brings us to the final, and most dangerous, consequence of the PLA's taking of Tiananmen Square: civil war.

Word of a split between group armies was already in the air on the evening of June 3, when citizens were telling incoming units that the 38th Group Army had revolted and that they too should follow this example and stand on the people's side against the government. By midday on Sunday, many believed that the 38th was positioning itself outside the city, preparing to launch an attack on the 27th Group Army. It seemed as though the conflict had shifted ground. It was no longer a battle between army and citizenry, but a battle between armies.

PLA soldiers themselves contributed to the notion of brewing conflict between units. Many people on Sunday encountered soldiers who expressed disapproval of how the Army was being used in Beijing. One soldier confided his disgust to a student at Muxidi. "I'm a PLA man," he said, "and I have to carry out my orders, even orders that might result in what you saw. Otherwise I'd be court-martialled. But let me tell you one thing: the ones who gave those orders are damned bastards." An eyewitness at the Jianguomen overpass who watched a military convoy drive out of the city at 10:30 heard an officer announce over a loudspeaker: "We the PLA did not fail you." As early as Sunday afternoon, some soldiers were seen wearing white armbands; on Monday, others wore red armbands. These armbands appeared to be used to distinguish certain units from others. Eventually, it seemed that every soldier was making an effort to show that he was part of the 38th. In the west end of the city, soldiers in a truck convoy flashed the number thirty-eight on their fingers to bystanders on the street.

The people of Beijing reacted to these signs in a way that reflected their own wishes. They judged that part of the PLA was disgusted with the way in which the Army had been directed to wreak violence on the demonstrators. PLA disaffection with martial law duty seemed to be confirmed when State Council spokesman Yuan Mu conceded at a press conference on Tuesday that four hundred PLA officers and soldiers had gone missing during the operation.

From the rumor of conflict between the 27th and other armies, notably the 38th, people projected good and evil: the good soldiers of the 38th had refused to engage in shooting citizens, whereas the evil troops of the 27th had committed all the atrocities. To this belief was added the hope that the 38th or another group army would take up the citizens' call for vengeance and wrest Beijing from the hands of the vicious 27th. Rumors on Sunday also had it that group armies from Fujian (the 31st) and Guangdong (the 42nd) were on their way to Beijing to challenge the city's occupation by the 27th. Although military transport planes are believed to have ferried troops of the 31st north to Beijing on Sunday and for the next several days, the relationship of these reinforcements to the 27th is a matter of speculation.

The idea that other group armies would punish the 27th involves several logical steps, some of which are difficult to substantiate. First of all, was the 27th Group Army, which quickly gained an evil reputation, the only unit involved in killing civilians? Soldiers of the 27th did kill some demonstrators, yet the killing went on in many different locations around the city. It seems clear, in fact, that the 38th was part of the column that fought its way through Muxidi and was as involved in the killing as the 27th. However much the people may have wanted to identify a scapegoat, the 27th was not the only unit with blood on its hands.

Second, some soldiers were unhappy with how the Army had been used, but can that personal unhappiness be translated into coordinated resistance on the part of their units? Individuals may have defected to the people's side, at least in spirit, but would any group army have dared at this point to mutiny against the Martial Law Command and the Li Peng regime? Mutiny involved enormous risks, particularly within an army such as the PLA, in which internal security supervision is intense. Disobeying orders means court-martial and automatic execution.

Orders were indeed disobeyed, and soldiers court-martialed. By good report, the site of the 1990 Asian Games, then under construction at the north end of the city, was used as a military execution ground in June. Rumor named the 40th as one of the group armies that was punished, purportedly because officers of the 40th planned to carry out a coup d'état against those who had sanctioned the Massacre and install a new regime to set things right. (It was the 40th that was said to have delivered rifles to Beishida.) This rumor may simply express what some citizens hoped would happen: that the PLA would redeem its honor by punishing the perpetrators of the Massacre and taking power from the hands of Li Peng and Yang Shangkun. More reliable, if less explicit, is the report of Yang's younger half-brother, Chief Political Commissar Yang Baibing, who announced at the end of 1989 that 111 officers had been punished for "breaching discipline" during the operation (21 of divisional rank, 36 of regimental and battalion rank, and 54 of company rank). Punishments were not reserved for officers alone; 1,400 enlisted men were also disciplined.

Although orders were disobeyed, it is difficult to believe in the theory of mutiny. The PLA tends not to play a role in Chinese politics independent of the Communist Party. Despite the reservations of many officers, including the head of the 38th Group Army, it is unreasonable to think that the PLA would split over suppression of the Democracy Movement. The Army prizes internal unity, and that unity is enforced through political supervision and stiff discipline. A split between factions in the Party may to some extent induce parallel splits within the leadership of the PLA, yet Army officers recognize the danger of backing particular factions or supporting contending ideological positions. Their predecessors had done just this two decades earlier during the Cultural Revolution, to the serious detriment of the PLA's reputation. However much individual officers and men disliked certain orders, it is unlikely that the PLA was vulnerable to taking sides.

The rumor of interunit conflict has to be explained in military, not political, terms. There was indeed outrage within the Army. Some of it had to do with dissatisfaction over being sent into an urban environment among civilians to deal with an essentially political problem, against the wishes of the people. But the main focus of outrage was not primarily over the suppression of the Democracy Movement. It was, rather, over the battle conduct of certain units. It appears that during the night assault on Beijing, a detachment composed primarily of soldiers from the 27th Group Army killed other soldiers, at least some of whom were members of the 38th. This was the key to the split.

The 38th formed the main core of the force that slowly fought its way in from the west through Muxidi and Fuxingmen before and after midnight. Contingents from other group armies were also included in this force—no force that night was composed of the soldiers of only one group army—but they appear to have been positioned behind the 38th. The 38th was given the tough task of clearing the demonstrators along Changan Boulevard. Its soldiers had to have been among those who shot demonstrators. But this force got slowed down. Discouraged by the sluggish progress of these troops, the Martial Law Command sent in another body of troops, composed primarily of soldiers of the 27th. These troops had orders to tolerate no obstacle and clear the way to the Square by a certain deadline. Whether through excessive zeal or gross miscalculation, the 27th pushed too hard. Rather than waiting for the troops in front of them to finish their task or move aside, at least one platoon of the 27th jumped the gun and either drove or gunned its way through soldiers and civilians alike. Soldiers belonging to the 38th, and possibly other units as well, were shot by the troops coming up behind them. The 27th was declared a rogue army.

I have no hard evidence for this reconstruction of events. It is based on secondhand reports from student leaders, journalists, and one foreign military observer. But it fits with so much else that we know. For instance, the English-language announcer on Radio Beijing offered the same report—minus the unit numbers—on Sunday, saying that "some armored vehicles even ran down infantrymen who hesitated to fight the civilians." The announcer cited eyewitness testimony for the allegation. Because Radio Beijing overlooks Changan Boulevard near Muxidi, the staff had good reason to know exactly what had happened. The story of soldiers killing soldiers is plausible. It was a new moon—visibility was low—and an AK-47 can kill at half a kilometer. The machine gun on an APC has an even longer range.

A NATO source has been quoted as supporting the view that some of the military deaths were from friendly fire. The killing of soldiers by soldiers would go a long way to account for the high casualties the Army suffered that night, which are otherwise inexplicable. The official count of fatalities on the government's side—soldiers and police—exceeds a hundred. This is an extraordinarily high number of deaths for an army engaged in fighting unarmed civilians. The NATO estimate of military casualties at first went as high as a thousand; this is improbable. But even if we accept the official Chinese figure, there is a major discrepancy between that number and the fewer than twenty who have been honored as "martyrs." One is left to assume that the others lost their lives in ways that cannot redound either to their personal credit or to the credit of the PLA.

The scale of troops and armor being moved into the city Sunday and Monday nights after the Democracy Movement was crushed thus began to make sense to Beijing residents. It appeared as though units were arming against each other. Stories of "good soldiers" preparing to do battle with "bad soldiers" swept the city. People were convinced that the city was under the control of the 27th, and that the 38th was planning to drive out the 27th and put Zhao Ziyang back in power. The rumor on Monday was civil war.

The Voice of America (VOA) picked up the rumor that conflict was brewing. To attest this rumor, VOA reported as well that heavy artillery fire was heard northwest of the city Monday afternoon. The report said that tanks had been seen rushing to battle on the western side of Beijing and conjectured that a battle was already in progress to the northwest. The BBC offered a more nuanced report, stressing that rumors of artillery fire northwest of the city were unconfirmed. Neither had eyewitnesses, but I do. Members of a foreign scientific delegation were out walking in the Old Summer Palace in the northwest suburb Monday afternoon when a thunderstorm blew up and they had to return to their dormitory. A member of the delegation (not an American) turned on the shortwave radio at 5:00 P.M. and was dismayed to hear the VOA turn a thunderstorm into artillery fire. "Damned old VOA always stripped everything conjectural out of their reports: everything they reported had happened. The broadcasters were implementing self-fulfilling prophesies, talking about chaos to people who were clearly on the edge of chaos. To say that something terrible had happened could cause the people who heard it to go over the edge and do something."

Armored movements late Monday afternoon gave the strongest sign that a split had occurred in the Army. Tank detachments and soldiers appeared at the Jianguomen and Fuxingmen overpasses, where Changan crosses the 2nd Ring Road, as well as at other strategic intersections along the ring roads. At both Jianguomen and Fuxingmen, the tanks were arrayed in an imposing defensive formation, their guns pointing outward. An observer watched the eastward detachment take over Jianguomen from the sentries posted there at about 6:00 P.M. "Seventeen tanks and a couple of APCs arrayed themselves in battle formation, facing in all directions. They weren't set up to fight off people with sticks or gasoline bombs. By this time we were hearing the rumors of the rogue 27th Army. We were told that the troops holding the intersection were from another group army. "When the tanks showed up, they pointed their guns right where these soldiers had parked their trucks."

After an exchange between commanders, the sentries withdrew to the center of the bridge and the tanks were arrayed around them. Another company of about 150 infantry, bivouacked three hundred meters north of the overpass, took over guard duty. To observers, the changing of the guard suggested that different units were involved. By best guess, the tanks were from the 1st Tank Division, which is affiliated with the 27th Group Army, whereas the sentries were from another group army (possibly the 16th or the 39th).

A total of twenty-six tanks and three APCs were eventually moved into position on the overpass. Most faced east, though a few trained their guns north and south along the 2nd Ring Road. The tank group's orientation suggests that it anticipated an attack from the east. And the decision to defend the overpass with tanks rather than foot soldiers indicates that the group's command was concerned about an attack involving more than rock slinging. The arrangement was in some ways peculiar for a defensive formation, because sitting on the overpass left the tanks exposed. But to set up proper tank defenses would have required bringing in bulldozers to excavate trenches where the tanks could dig in, leaving only their turrets and guns protruding above ground. This was not feasible on a concrete overpass. Nor was the Army in a position to be destroying the buildings around the intersection and using them as screens. The buildings on the northeast corner, after all, were diplomatic apartments. One has to conclude that the tanks were deployed as well as they could be in the situation.

The tanks remained in position on the Jianguomen overpass for the next two days. The only challenge thrown against them came on Monday night. It was the lone voice of a foreigner angered by this show of military force below his apartment. Invoking Lei Feng, a much belabored PLA model of self-sacrifice from the 1960s, he yelled the once-popular slogan "Study Lei Feng!" out his window several times, ruefully adding that "that could be why they shot my window." Otherwise, the overpass remained quiet. A separate armored column was parked a kilometer further east on Changan Boulevard on Monday night, though its relationship to the tanks on the bridge is unknown.

That night, the sound of sporadic gunfire continued to echo around the city, especially out in the suburbs. The shooting seemed to confirm popular suspicions that the Army was at war with itself, as a Chinese student argued to me later. "There is no way this could have been conflict between soldiers and Beijing civilians. Soldiers can have guns, civilians can't. People with guns were firing because there were other people with guns." Another student in exile gave me the same reaction. "People guessed that there was conflict simply by the sound of gunfire every night out in the suburbs. You could only have that sort of fierce fire between military units. How could ordinary people go out there and fight them when they had no weapons? Sometimes also you heard heavier explosions, like cannons, or maybe grenades. We didn't know what they were. You didn't hear it in the daytime, only at night." The explosive booms were short and dry in sound, sometimes two or three in quick succession. "We heard what we thought was shellfire," reported a foreigner on the east side of the city. "It was to the south of us, and it lasted long enough that people started running around saying, 'Did you hear that? What was that?' Everyone was convinced it was the sound of shells being fired, though we all kept hoping it was thunder." Another group of foreigners in the northwest of the city concluded the worst. "We could hear heavy gunfire—tank or artillery or some such—in the distant outskirts northwest and southwest of the city. We all discussed it amongst ourselves and decided we were not hearing thunder." An American student at People's University who heard the same sounds on Monday night was told next morning that "the 27th Army shot at the unit coming in, not necessarily to hit anything but make it clear that they were not allowed in the city."

The story the American student picked up was close to the truth. A Western military analyst living in the eastern part of the city recognized the explosions as the sound signature of tank guns, not artillery. When artillery is fired, there are two booms, one when the shell leaves the gun and another louder boom when it hits its target and explodes. Unlike artillery, a tank does not fire an explosive round. When a tank gun is fired, there is usually only one boom, a terrible noise accompanied by a ball of fire. There is no second boom, because a tank usually fires a hard metal rod designed to penetrate the armor of another tank. What people heard that night were tank probes. This is a tactic an army uses to test the location and size of posts a defending army will put up around the perimeter of the region it controls. The challenging army sends in a vehicle, usually an APC, along an axis road until it is engaged. The purpose is to find out where the enemy's outer defenses are, and what armor they have deployed there. When the enemy's tanks detect the incoming vehicle, they fire at it. Two or three tanks may fire at the same time, creating the impression of one gun firing several times in quick succession. As soon as the vehicle is engaged, it speedily withdraws and reports back.

This kind of probing action went on Monday night on the main north, west, and south axes at least ten kilometers beyond the 3rd Ring Road. Only the eastern axis remained silent. Whether any greater conflict occurred that night can only be guessed. A U.S. marine claimed to have seen a damaged armored vehicle in the eastern part of the city that had been disabled by an armor-piercing shell. The next day, a BBC news team in the outskirts declared it had found evidence that weapons battles had been waged. The BBC and VOA also reported conflict between units in the south around Nanyuan Airfield. But we have no eyewitnesses.

We do have a soldier's testimony, however. On Tuesday, a PLA officer called a longtime foreign friend and confirmed that armies had been squaring off against the 27th. He said that the 38th was giving up its struggle against the 27th, but that the 28th from Shanxi was preparing to take over the fight. He was quite certain that civil war loomed. The officer added that other units were still smarting from the assault Saturday night. Apparently, when troops of the 54th from Henan tried to enter Beijing from the southwest, they had refused to shoot civilians, and "they'd had the hell beaten out of them because they didn't want to fire." The 54th was under a cloud, restricted to its temporary base in the PLA's August First Film Studio in the southwest suburb.

Was armed civil war on the horizon? Was the 27th Army really about to turn the Beijing Hotel into a military headquarters? Was the 38th really going to close the Beijing Airport in order to airlift troops to fight on the students' side? Rumors flew thick and fast Monday and Tuesday. A foreign teacher told me later that he was skeptical of the assertion that China was on the brink of civil war. It was a case of "wishful thinking," he said. Everyone wanted to see the PLA punish perpetrators of the Massacre. As much as they feared civil war, the citizens of Beijing wanted justice, and this seemed the only hope of getting it. The PLA was too professional, he insisted, and too securely subordinated to civilian leaders, to get involved in a coup.

The problem here is that "coup" and "civil war" are not the right words to describe what was afoot. Much as segments of the Army disliked the order to enforce martial law in Beijing, individual units did not have the power to challenge their orders. And much as segments of the Party leadership shared that dislike, the core of that leadership was secure. Zhao Ziyang was well outside Deng's confidence by June 3 and did not have the political clout to dispute martial law. The PLA would not dare back him. The Party's decision later not to lay charges against Zhao, nor strip him of Party membership, indicates that he was never part of a military challenge against Li Peng or Deng Xiaoping. China was not on the brink of civil war in the sense of contenders for power fighting for supremacy. The Army was not engineering a coup. But China was on the brink of an internal military conflict that threatened to destabilize the regime.

The big puzzle is why the political leadership did not bring the Army to heel as soon as signs of conflict showed. Only three conclusions are possible: the political leaders were ignorant of the conflict; the military leaders refused to listen to their political masters; or the latter were too absorbed with more pressing problems within the Party. The first conclusion is impossible, the second unlikely. The third seems to have been the case. For the first few days after June 4, the political structure lost unity at the top. The consensus it needed to proceed with the suppression was not there. State Council spokesman Yuan Mu did appear on television on Tuesday to say the government supported what the Army had done. But the lights in the Great Hall of the People burned through the night, and the media on Wednesday gave no clear sign that the political problems were resolved.

A spokesman for the Ministry of Foreign Affairs issued a statement that day saying that the government was stable and in control of the situation. But officials inside the State Council conceded to an interviewee

that factional struggle was going on, and that they did not know its outcome. What particularly caught the interviewee's attention was the fact
that the State Council was running out of gas for its cars. His conclusion:
the State Council and the Army were not working together.

Bureaucrats who could be reached that week conveyed the same sense
of the powerlessness as the State Council. When four Western ambassadors called on the Ministry of Foreign Affairs on Thursday afternoon
and complained that soldiers had fired on foreign nationals the day before,
a Ministry spokesman said that it was impossible for Foreign Affairs to
guarantee the security of diplomatic staff at night time because of night
engagements between soldiers and "ruffians." He asked that foreigners not
go out at night for their own safety. In essence, the foreign community was
being told to submit to a self-imposed curfew. When asked whether China
would respect the Vienna Convention, the spokesman could say only that
he would see what could be done. The effect of his comments was to
suggest that the civil government was not in power.

The incident the ambassadors were complaining of—and the final sign
of interunit conflict—occurred the day before, on Wednesday shortly before noon. A column of troops driving east out of the Square along
Changan Boulevard opened fire at Jianguomen. A foreign businessman
came out the side entrance of an office building just beyond the overpass to
see

> open troop carriers, each one full of soldiers holding automatic weapons
> ready in their hands. On one vehicle the soldiers were shouting slogans:
> "Love the citizens of the capital! Protect the citizens of the capital!" I waited
> at the light for the convoy to pass when from the direction of the Jianguomen
> overpass we heard sudden gunfire. I ran back down the alley. The firing
> started at the overpass and moved right past us. Soldiers on every truck were
> shooting, as well as the soldiers walking beside the convoy. The automatic
> weapons fire sounded like rain on a tin roof. I could see a lot of bullets
> bouncing off the China International Trade and Investment Corporation
> (CITIC) building across the street. Two helicopters moved in as soon as they
> started firing, passing very low, but there was no shooting from them. The
> shooting lasted for five minutes and didn't stop until the convoy had gone
> completely past.
>
> From what I could see they shot only at buildings. I went straight north
> as fast as I could. When I got past the CITIC building, I heard gunfire break
> out again from further east. I heard later from an architect working on the
> new complex at the Dabeiyao intersection that the hotel there lost ninety-five
> sheets of plate glass. That night the news report was that several hundred
> rounds were fired at a sniper in a diplomatic apartment building—but more
> than several hundred rounds were fired during that trip down the street.

Why did this convoy let off this destructive *joie de feu* when, by
Wednesday, nobody dared stand up against the soldiers? The sniper theory is not persuasive. The standard response to a sniper threat is to spread

out, but the soldiers around the trucks continued to move in a tight formation less than a meter from one another. Furthermore, when troops went into the diplomatic compound and removed a Chinese suspect, they did not treat him roughly, as they would have treated a sniper. There is the further problem that none of the foreigners in the compound detected any firing from their buildings. So what was the sniper story intended to cover?

Some foreigners assumed that the purpose of shooting up the Jianguomenwai diplomatic compound was to intimidate them. This explanation neglects the fact that, although many buildings were hit along this stretch of road, the firing targeted one part of one building: the third and fourth floors of Building No. 1. Bullet damage reveals further that the fire came not just from AK-47s in the street below but from a light machine gun in the recently completed hotel across the street. Troops had been seen entering this building the day before, and they had removed windows on the fourth floor, presumably to set up a firing position. Despite appearances, the firing was not random. What, then, were they shooting at?

It so happens that at least two foreign television crews were on the third and fourth floors of Building No. 1, which gave them an excellent position for shooting footage of the tank movements on the Jianguomen overpass. One of them, a crew from Canadian Television (CTV), had been there since the first column of troops was stopped Saturday evening and had been shooting the Army's actions on the overpass ever since. After most foreigners had vacated the building on Monday night and Tuesday, these crews stayed on and continued filming. Their videotapes they pigeoned out with air travelers flying to Hong Kong and Tokyo, and the images they captured were relayed by satellite to North America and shown on television. Troops came into the diplomatic compound with bullhorns on Tuesday to announce in Chinese that any foreign journalist seen taking pictures in contravention of the martial law regulations would be shot at. An ABC crew filming surreptitiously from a balcony of the Beijing Hotel had already been fired on by a troop convoy that spotted the camera. But foreign cameramen in Building No. 1 continued to defy the order.

The Army needed to stop these cameramen from broadcasting the size and location of this formation. The concern was not with the uncomplimentary coverage China was getting from the international media. It was linked, rather, with the threat of interunit conflict. The unit holding this overpass did not want the location, strength, and positions of tanks on the Jianguomen overpass televised for all to see. This was strategic information that another unit contemplating an attack could make good use of. The foreign cameramen were unwitting pawns in other people's battle, giving free military intelligence to the other side. Shooting up Building No. 1 under the cover of random fire from a troop convoy was a counterintelligence operation. It worked. The journalists left.

This incident was the last sign of interunit conflict. By late Wednesday, the conflict came to an end off the battlefield. The Party reestablished its unity and reasserted its control over the Army. When Premier Li Peng and Vice-President Wang Zhen appeared on television Thursday to congratulate martial law commanders in the Great Hall of the People, the ambiguous "civil war" was over.

Although PLA units continued to enter the city, Thursday marked the beginning of a return to order. Wrecked vehicles were towed away. Traffic cops reappeared at some of the intersections where soldiers weren't posted, though the gas shortage had reduced traffic to a trickle. Streets were swept. The PLA was a conspicuous part of the cleanup. "Soldiers with straw brooms were out sweeping the streets, though two soldiers carrying weapons were posted with every group, one in front and one behind. There were also trucks going around with loudspeakers blaring out messages to people."

Only on Friday, June 9, did this week of anxious waiting finally come to an end. Deng Xiaoping appeared at dinnertime on television, congratulating his military officers for their courageous and disciplined work. A foreign observer registered the change that evening. "When I went back out into the streets around 7:30, I sensed a sigh of relief. The fear of civil war was already abating now that they knew Deng was alive. I talked to a couple who said they felt better now that 'the emperor had reappeared.' The man said he could go to his factory and work harder. When I asked why, he said, 'Now I have to produce more so that there will be more for me.' Economic concerns reappeared very quickly."

With military security reestablished, the long process of arrest and repression could now begin. This process would go on until the last trials of leading activists and participants two years later. It started on Wednesday, when the evening news called on the people to turn in protesters. Viewers were told to dial 512–4848 or 512–5666, and were assured complete confidentiality for whatever information they might supply. An accused would have no chance to challenge his accuser. These hot lines remained open for several days, although Chinese abroad did their best to jam the lines with an uninterrupted flow of incoming international calls. The arrests started that weekend, as police began to round up the organizers of the Workers' Autonomous Union. The repression was in place. Everyone was going back to work. One could do nothing else.

A foreigner received a telephone call Monday from an acquaintance who worked in the Ministry of Foreign Affairs. They had arranged to meet in a park, but she called to cancel.

"The weather has changed," she said. "It's raining out, you know." The sun was shining.

8

Closing the Century

The state has been a conspicuous piece in the puzzle of how China has responded to the challenges of the twentieth century. The frail imperial state structure collapsed in 1911 when the Qing empire fell, and the republicans who pushed it aside could not build an effective replacement. Out of the shifting morass of warlordism arose two Leninist state-organizations, the Nationalist Party under Chiang Kaishek and the Communist Party under Mao Zedong. Both constructed political machines designed to funnel public opinion into a unitary consensus supporting their policies. But military force was decisive in both cases to achieve and sustain national supremacy. The state on its own was powerless.

After winning power in 1949, the Communist Party did not sponsor the creation of a new state, but became it. On the people's behalf—and for their own good—the Party monopolized all political power and exercised totalistic control. It "served the people," but without consulting them. The Party's monopoly has prevented the Chinese state from existing as a forum for representing and negotiating competing ideas and interests. Demands arising outside the Party have no legitimate sphere in which to find expression. Organizations autonomous from the Party have no legal basis on which to stand. In case of conflict, final arbitration is handled by the Army.

The Democracy Movement arose in the gap between the unitary idea of a Party-dominated state and the multiple reality of social existence. The gap itself threatened the Party's grip on power, yet the Party could have managed the Movement in ways other than it did. The Beijing Massacre was not the only possible outcome of the growing tension between state and society, nor did it have to happen. It occurred for the specific reason that the Party as a state had available to it military means and chose to direct those means against the civilian population of Beijing.

The basic fact of the Massacre—the state's use of lethal force against civilians—is the point from which all interpretation of the event must flow. From it too flows this chapter, in which I work outward from the decision to use lethal force toward the three questions that have came to dominate my thinking about the military suppression of the Democracy Movement. The first question regards the Chinese government: What were its intentions in mobilizing the Army against civilians? Did it intend that the suppression be violent? The second concerns the Army itself: Why did the troops act as they did, and what does this say about how the PLA might act the next time around? The third question is abstract; it asks about the larger context in which the suppression occurred. To put it in question form: How does the military suppression of the Democracy Movement help us understand China's predicament toward the close of the twentieth century?

When Deng Xiaoping appeared on television on June 9 to praise his martial law commanders, he insisted that the suppression had altered nothing fundamental in the Chinese government's course. The policies of reform would continue. Despite the international condemnation, he vowed that China's leadership would "never change China back into a closed country." Yet by mocking international concern, as he did, Deng seemed to close the very doors of the late twentieth century he himself had opened. It was the last decisive act of his political career. Short of a major upheaval in the coming decade—possibly the death of Deng himself—it may also prove to have been China's last significant political decision for the rest of the century.

Deng was not disconcerted by the destruction or the response around the world. "The storm was bound to happen sooner or later," he declared. "It was just a matter of time and scale." His words project an image of supreme confidence, of a leader who enjoys full control of his government and his military. They also suggest that Deng regarded both the Democracy Movement and its suppression as inevitable. It is a convention of Chinese Communist rhetoric to assert that whatever happens had to happen. Deng may have been doing nothing more than wrapping himself in the mantle of Communist wisdom. Even so, the remark implies that the Massacre was a desired outcome. Deng would have us believe that the operation went as planned; he offered no sign of contrition or regret.

Certainly, the scale of the operation, involving at the very least 150,000 soldiers, along with substantial armored support, testifies to considerable planning. The first order bringing outside troops into the Beijing Military

District was issued before April 25. Military leaves were cancelled by the beginning of May. At least two-thirds of the final force was in position for the first assault on May 19. From the first, Deng considered the Army an appropriate means for stopping the students' protests. He sided with it to the last.

The delay in bringing military means to bear on the Movement does suggest a willingness to try other means first, though always with the understanding that if a political solution failed, a military solution would follow. Some of the delay, though, stemmed from the character of the force Deng wanted assembled. He was not content to use the group armies stationed locally within the Beijing Military District, even the Region. In the end, units from eighteen provinces and two municipalities came to Beijing to be part of the martial law forces. Deng wanted wide participation in the suppression, perhaps because he knew it would be an unpopular action, both among the people and within the Army. Arranging that participation required time to bring the armies into position.

However much some commanders resisted having the PLA used to discipline a civilian population, the Army was not a major player in the decisions about when or how to bring the Democracy Movement to a close. The 38th Group Army command dragged its feet prior to the first offensive. But that was only one unit, not the entire PLA. And it did in the end take part in the offensive, under different command. The idea that the PLA resisted Party commands presumes a separation between Army and Party that goes against the principles by which power is organized in China. The Party rules everything, and the military command has been structured in such a way as to foreclose any other possibility. In part this stems from the Party's history of having been founded at a time of warlordism. Such a Party had to learn to survive long years of persecution, which meant having to acquire military power. With the widely discussed exception of General Xu Jingxian, who it is believed was willing to face court-martial for his refusal to lead the 38th, PLA commanders understood that they had to come on side. Their job was not to adjudicate competing political claims between the Party and its opposition. It was to obey the Party's orders.

The failure of the first offensive meant further delay while more troops gathered outside the city and soldiers were infiltrated into the downtown core. This delay fueled wrong expectations: that the Army was powerless against the citizenry, that the soldiers sympathized with the students, that the Party leadership did not have the political will to use violence against civilians. These expectations meant that few could believe on the evening of June 3 that the troops would actually open fire. As late as the early hours of Sunday morning, recalls a Beishida graduate, he and his friends "figured the worst that could come of this would be a wave of arrests, aimed particularly at the student leaders." They could not imagine slaughter. Another Beishida resident presumed the same thing that night. "We knew that something was going to happen. We thought they would move in, use tear gas, rough them up a little—or a lot—and throw them in jail. We never expected that they would go in there shooting."

Although delay may have jeopardized caution, it also gave the Democracy Movement time to keep growing. The presence of large numbers of protesters, standing now in the shadow of the Goddess of Democracy, could only have increased the irritation of the Army's political masters. As one foreign eyewitness observed to me four months later, "The longer the wait went on, the more humiliating it became for the government to have the Square taken over, not just with demonstrators but with riff-raff and garbage. Tiananmen became a people's square, and every day that went on, the need to clear it must have grown greater and greater. And so, desperate tactics were resorted to." This however is pure supposition on his part. It presumes that the suppression was contingent on how the Movement developed. As I argued in Chapter 4, this way of thinking is problematic. In retrospect, all the signs indicate that the second invasion of the city would go ahead regardless of what happened in the Square. Still, the Goddess of Democracy and the final seventy-two-hour hunger strike did not open any ground for compromise.

By frustrating the leadership and feeding the confidence of the citizenry, the delay may have contributed to the violence on the night of June 3. The government may not have anticipated the scale to which it grew, however. Even Deng Xiaoping conceded that the scale of "the storm" could not be predicted. If there is any sign that the outcome was out of proportion to government expectations, it is the uneasiness over the scale of casualties. Official propaganda has consistently set casualty figures low. The official death toll of three hundred is flatly contradicted by hospital estimates given in Chapter 6. No government likes to fully tally its victims, but the regime's unwillingness to substantiate its casualty figure suggests to me not just that the death toll was high but that it was higher than even this regime considered appropriate. The official numbers may indicate what the government considered acceptable and what the Army failed to deliver.

There are other indications that the government did not fully anticipate the destruction on the night of June 3. One foreign reporter thought to judge the degree of premeditation by seeing how the television news at 8:00 A.M. Sunday morning handled the incident. "The announcement was ready for broadcast, yet it was put together without pictures. The announcers were wearing all black [a gesture that would be forbidden later that day]. They looked like they were going to cry. We saw their faces only briefly, then it became voice-over. I can't understand why they didn't have any visual footage. If they had planned the Massacre, they would have had the whole thing ready and used military announcers. The first news is important, and they screwed it up."

The lack of preparedness in hospitals points to a similar failure to anticipate the scale of violence that night. As noted in Chapter 6, the rumors that hospitals had advance warning of the slaughter are contradicted by the obvious inability of the hospitals to accommodate the casualties brought to them. If they did have warning, what they were warned of was not equal to what they had to cope with.

The government could not hope to control everything that happened in the streets. Even government agitators would have found it difficult to direct the crowds—and foreign military observers did not detect their intervention during the heat of battle. It is impossible in any case to know how crowds will react in crisis situations. Nonetheless, by dispatching large numbers of armed soldiers among civilians, the Chinese government set up certain probabilities about what might happen. Later, official spokesmen would cling to the view that the operation was designed to restore order to a city gone out of control. This is the theme of televised remarks made by Jiang Zemin, the new Party secretary general, on May 19, 1990, a year after the imposition of martial law. "I don't have any regret about the way in which we dealt with the events which took place last year in Beijing," he said. "Had we failed in the end to take resolute measures to deal with those events, then the entire capital of the People's Republic of China would have been thrown into great chaos." Those measures, he said, China's leaders "were forced to take." He further insisted that they were within the normal range of response that could be expected from "any government in the world." In his view, the military operation on the night of June 3 was a matter of reimposing law and order, not of stifling public debate with violence.

Given the plan of attack that night, it is impossible to maintain that killing was not part of the plan, or that the troops had no other choice than to shed blood. The ferocity of the assault signals that, once the order to fire guns was given, this operation was designed—or, at the very least, expected—to cause violence and produce civilian casualties. No government sends tens of thousands of combat troops into a volatile urban setting without expecting violence to erupt. Pointing AK-47s at unarmed civilians is a formula for disaster.

In regard to the government's intention, a pivotal point in the chain of events Saturday night was the order to open fire. The first assault on May 19 was conducted without shots being fired. During this first operation, some units carried assault rifles, but they were not loaded. The Army distributed ammunition to some units during the night of May 21, presumably in consultation with Party leaders, but permission to use it did not come down. Violent assault against civilians was not yet part of the plan. It may have been reserved as a contingency, but the operation was aborted short of reaching that stage. The government's unwillingness to authorize violence during the first assault strengthened the students' conviction that there was safety in numbers, that, as a foreign eyewitness put it, "the more people there were in the Square, the less likely they'd get suppressed. They assumed that the Army wasn't going to go in and mow down tens of thousands of people."

By 11:00 P.M. on the night of June 3, soldiers were shooting civilians at Muxidi. They would continue to do so there and elsewhere in Beijing for several days. As long as the soldiers had loaded weapons, they carried with them the option of firing on civilians. The Martial Law Command would not also have included teams of marksmen among the troops had it not anticipated a use for them. On Saturday night, the question was at what point, not whether, live ammunition would be authorized for use.

The Chinese government has engaged in extensive propaganda work to demonstrate the necessity for the decision to unleash lethal force. Using videotapes from the British-made traffic monitors installed at major intersections, as well as the footage shot by PLA videographers, the government has produced television documentaries that purport to show how brutally protesters were attacking the soldiers. By reversing sequence and showing riotous behavior before the scenes of troops firing, these documentaries convey the impression that the demonstrators turned to violence first, and that the troops adopted lethal measures only after the relentless attacks of vicious hooligans. In fact, most of the "rioting" footage was shot *after*—and in response to—violent acts committed by the Army. We see protesters bombarding troops with rocks without being told that the troops had already opened fire. We see crowds attacking stalled APCs without being shown APCs killing and injuring both civilians and soldiers. Sound has been manipulated as well as sight. We see bystanders hysterical with anger but do not hear the sound of automatic weapons fire that has been edited out of the soundtrack.

The naive viewer is led to the conclusion that the government had no choice but to quell the turmoil in the streets. As the voice-over on one video phrased it: "China would surely have been thrown into serious turmoil, and the people's republic would likely have been subverted. Compelled by the circumstances, the government had no alternative but to declare martial law in parts of Beijing."

Within Chinese public opinion, there has grown some sympathy for the government's assertion that the protesters started the violence by rioting and burning vehicles, and only then did the soldiers respond with lethal force. A Chinese student I interviewed a year after the Massacre told me that people in Beijing were discussing this matter. "Did the soldiers open fire first, or did the students set fire to their vehicles first? There's a lot of debate among the people about this. Some, like my parents, say that the students set vehicles on fire, and only thereafter did the troops open fire." (A Tibetan critic has argued, persuasively in my opinion, that the Chinese people are vulnerable to government appeals for the need to use violence. "When the Chinese government repressed its minorities, the whole nation cheered, offering enthusiastic and firm support," Jigme

Ngapo has pointed out in a commentary on the Massacre. "Tolerance of violence that goes as far as to condone it has actually encouraged this regime to increase its use of violence.")

Participants in the night's events knew that the citizens did not start the violence. Opening fire was not simply a contingency option that was brought into play once things got out of hand. As the Chinese student I just quoted went on to note, "The Army Command was in communication with its troops and their attack was well organized. So my feeling is that the Army opened fire first. Why would we students start it? Before, our relations with the soldiers were good, very friendly. Why would we turn around and burn their vehicles?"

There was simply too much killing, and it went on for too many days, for slaughter not to have been built into the plans for the June 3 assault. Its soldiers having been turned back once, the Party could permit no second defeat.

But it was not just pride and vengeance that drove the Party to conduct the slaughter. Soldiers had to open fire in order to destroy the citizens' sole protection: nonviolence. The government could draw on no political resources to achieve this. Nonviolence created a charmed circle that protected the Movement against the persuasive and coercive mechanisms the Communist Party is usually able to employ. By resorting to violence, the soldiers stood a good chance of inciting violent responses from the people. Only then would the government have a plausible pretext for striking down those who disobeyed its orders. A foreign diplomat in Beijing agrees. "As long as the demonstrations were absolutely peaceful, there was little the government could do. They had to provoke violence and riot in order to justify their actions."

Violence was accepted by the Chinese government because the Movement had to be suppressed—and no other means were available to a party that had chosen to close off all possibility of negotiating political claims. To permit the Movement to continue would, in the words of President Yang Shangkun on May 24, have led to the transformation of socialism into liberalism. "This would mean the destruction, over-night, of the People's Republic and of the achievements that are the result of decades of war and the blood that was shed by tens of thousands of revolutionary martyrs. It would mean denial of the Communist Party of China." The founding generation of Chinese Communism could not bear such a fate. "To yield," admitted Yang, "would mean our end." He was probably right. As Deng Xiaoping said, the only two things in question were time and scale.

The conduct of the Army on the night of June 3 reinforces the view that the government intended to use violence to crush the Democracy Movement. There is no reason to level assault weapons against unarmed civilians unless you intend to kill them. Even if you accept that troops are needed to disperse political opposition that will not succumb to persuasion, live ammunition will not do the job. It will only kill whoever is in range. As one foreign military analyst in Beijing put it to me in no uncertain terms, "That is the part that is revolting."

The martial law troops used a wide range of weapons against the populace. The AK-47 automatic assault rifle was most common. Soldiers also had metal bars, nail-studded clubs, garottes and whips, including a type of steel-core whip with an outer rubber covering that leaves no visible wounds. In addition to antipersonnel weapons, the Army used machine guns designed to be fired at vehicles and aircraft. Lethal weapons were made more lethal by being placed in the hands of incompetent and badly officered soldiers. Levels of troop ignorance were high. Soldiers did not show that they had much sense of how the operation was to unfold. Each seemed to know at most only his part, and did not understand how his part fit into the larger plan. As a result, soldiers became passive when the order they had received could not be carried out. They had no backup plans, and were not encouraged to take initiatives if a plan failed. When the driver of a stalled PLA truck hopped out of the cab on East Changan Boulevard early Sunday morning to get his truck going again, he was stoned to death by angry bystanders. None of the soldiers in the other trucks in the long convoy did anything to assist him: rescuing a fellow soldier was not part of the plan.

PLA accounts of soldiers' conduct make much of their suffering at the hands of wild mobs. They seek to establish several points: that the soldiers were under strict orders not to injure any innocent people, that they undertook punitive actions only against rioters, and that they practiced forbearance in the face of savage attacks. Some soldiers that night may have been animated by a subjective commitment to sacrifice themselves before harming the people, but their subjectivity is overwhelmed by the objective situation into which their leaders placed them. Soldiers were brutally attacked by enraged mobs, but that does not put the Army in the right. It was the Army that precipitated violent crowd reactions by sending armed soldiers into situations they could not control. Individual acts of heroism and restraint occurred but cannot redeem the Army as a whole. Soldiers, as is always the case, were pawns suffering for political ends that made a mockery of the Army's professed ideals of upright, professional conduct.

The composition of the martial law force exacerbated the confusion—and possibly the indifference of some soldiers toward others in difficulty. The Martial Law Command chose not to preserve existing units and send in soldiers who shared a common command and might identify with one another. Rather, it mixed subunits from different group armies within troop columns. This composition of forces made the entire PLA, and not one group army, responsible for the suppression. It also reduced the chance that officers would resist the order to occupy Beijing. It inserted them into a new command structure that overrode preexisting officer-corps solidarities within group armies.

This mixing of units complicated the problem of direction in the field. In crisis situations, group army loyalties pulled against interunit cooperation. There are hints of this tension in the battle reminiscence literature produced after the Massacre. One piece by a colonel in a communications unit belonging to the Beijing Garrison dramatizes the tension by showing him mulling over whether to guarantee the communications needs of his own unit or jeopardize them to help another group army. His staff officer is stunned when he declares, "We must protect the army of our friends, no matter what." A divisional commander in the other group army later praises the colonel for his unselfish contribution to his unit's success. The incident was deemed worth recording because it was exceptional.

To raise yet another coordination problem, the failure of provisioning is puzzling. Every general knows that a war can be won on the battlefield but lost in the supply lines. The PLA was never in danger of losing its war with the citizens of Beijing, but the soldiers in the Square had almost nothing to eat from Sunday to Tuesday. It was not until midnight Tuesday that the Army was able to move into the city its first major supply convoy, forty-nine trucks protected by five tanks, five APCs, and six hundred soldiers. The Martial Law forces ran out of other things besides food. On Tuesday, some units were sending soldiers out to scour local shops for batteries. How could the quartermastering for the martial law forces have been so shoddy?

Incompetence and poor planning may explain some of the chaos and account for some of the casualties on the night of June 3–4, but ill preparation and unintended mishap do not excuse the violence. Despite the gross errors, the operation as a whole went roughly as intended, and the soldiers acted largely as they were directed to act. A student participant has come to the same conclusion. "The soldiers seemed to be under pretty good control throughout the whole operation," he had to admit. "They followed their orders. Reports afterward suggest they were very poorly trained, shooting some of their own people, running down some of their own people. Maybe the operation wasn't well organized. But the soldiers followed orders."

It was the shooting and running down of other soldiers that threatened to turn the suppression of the Democacy Movement into a civil war. As PLA units shifted their attention from the protesters to a rogue army, coordination and discipline became even more difficult to maintain. Widespread rumors of military executions in mid-June hint at the extent to which control was lost. By the end of the year, according to the PLA, 111 officers and 1,400 enlisted men had been disciplined. And in May 1990, the commanders and commissars of six of the seven military regions were relieved of office or transferred to other posts.

In addition to the split that developed within the martial law forces, the PLA also suffers now from a split between those who took part in the suppression and those who did not. An Army officer whose unit had not been part of the martial law forces was at Beidaihe later that summer. Beidaihe, a resort on the ocean 250 kilometers east of Beijing, is the most exclusive summer retreat in China. Party and Army leaders go there for their summer holidays. This officer learned that officers from the 64th Group Army who had taken part in the suppression were also there. He was never allowed to speak to them. They were kept segregated, in total isolation. The PLA is an army in deep division.

Participation in martial law is now regarded inside the Army as a matter of shame. If asked, every PLA officer and soldier will insist that *his* unit was not involved in the Massacre. When a representative of the National Education Commission was lunching with officers at the National Defense University in November 1989, every one of them was adamant that his unit did not have the people's blood on its hands.

At best, lingering tensions within the PLA will undermine the Army's ability to coordinate operations between group armies in the future. The 27th continues to be a pariah. At worst, factions within the Party may build bridges with factions within the Army in the hope of outmaneuvering other factions in political contests in the future. And factions within the Army may now be more willing to be drawn into political contests than they were when unity was still maintained.

In the end, I have to conclude that, although the government intended that the suppression be violent, much of the destruction of life and property between June 3 and June 6 was unintended and resulted from unprofessional conduct on the part of the Army. PLA soldiers functioned poorly under the circumstances both because they lacked training in urban warfare and because their deployment in the field was inadequately coordinated at the command level. The startlingly high level of military casualties—a level that the dissension within and between units only made worse—confirms this impression of incompetence.

However badly the PLA troops performed, the fact remains that the military played the pivotal role in determining the outcome of the struggle between the Communist Party and its opposition in the spring of 1989. It is difficult to imagine how the power of Deng Xiaoping and Li Peng could have been shored up without the Army's intervention, though what the outcome might have been had the Army remained in barracks is anyone's guess. The actual outcome—both politically and in terms of the reimposition of civil order—was as a direct result of the military's involvement, however incompetently managed. The regime as it stood after June 4 existed by dint of armed force.

The decision to follow orders and rescue Deng from his critics was a fateful one for the Chinese military. By carrying out Party orders and allowing itself to be mobilized against the street demonstrators in Beijing, the military sided with the ruling elite. It did so at tremendous cost, in terms of both internal demoralization and popular disillusionment about the PLA's being a "people's" army. Even so, the PLA's intervention on the Party's side was to be expected. This is after all a Party army. The military's thorough subordination to the Party is fortified by indoctrination (through its political commissariat), intense supervision, and a recognition of shared material interests. The real surprise that spring was not what the Army was capable of doing, but what the Party would order it to do.

Still, it is not enough to say that the Army's role in suppressing the Democracy Movement was a function of automatic loyalty to political masters and leave it at that. The relationship between army and state is much more interesting, and more ambiguous. No military, the Chinese included, intervenes as a neutral force in a domestic political crisis. Politics always impinges. Even when orders are being obeyed, sides are being taken. This was certainly true in the spring of 1989, for just as the Democracy Movement forced sharp cleavages within the Party, so too it divided the Army. It thus behooves us to consider how the PLA leadership regarded the demands of the Democracy Movement, and how it might react to similar demands in the future.

The history of Third World praetorian regimes in the twentieth century suggests that the military has tended to intervene in domestic politics in favor of the new elites that have emerged during the process of "modernization" to replace nineteenth-century oligarchies, first to help them overthrow the political power of the old ruling classes, then to protect them against incursions from below. The sharing of interests between the military and the new "modernizing" elites is not surprising. The military is itself a "modern" institution, bureaucratic in structure and favoring the sort of rapid economic development that will enable it either to produce or purchase elsewhere the increasingly expensive hardware of modern warfare. The officer corps of Third World militaries tend to be filled, furthermore, by the sons of the new urban middle classes. These young men increasingly turn to the military in search of advancement at a time when status patterns are changing and economic crises make entry into other professions unattractive. In addition, the development of the technology of modern warfare has strengthened this trend: as weapons become more sophisticated, armies need to recruit college graduates to design, maintain, and operate them. And with greater sophistication comes greater cost, which can be borne only by embarking on ambitious programs for economic growth. To an ever-increasing extent, then, military officers and modernizing elites share common social origins that encourage them to embrace the same modernization goals.

The PLA is an exception to this model only to the extent that the formation of a modern military in China has been a two-stage process. During the revolution, the modernizing elites of the Communist Party relied heavily on peasant mobilization to achieve their political and military objectives against the old ruling powers. The original Red Army was officered as often by peasants as by the new middle classes, but it was a modern-style institution involving professional training and routinized structures of command. Committed to overthrowing the ancien régime and comprador elites that the Party identified as the obstacles to the building of a modern nation, the Red Army fought a successful civil war that led to the founding of the People's Republic in 1949. Mao Zedong, a member of the emerging middle class struggling to create the new order (he may have been born to a peasant family but he graduated from teachers' college), blocked the full professionalization of the military that his leading generals desires. He did so to prevent it from slipping from his control. The Army remained self-consciously peasant-proletarian as long as Mao remained in power. As soon as he was gone, however, Deng Xiaoping (with the help of Zhao Ziyang) initiated reforms designed to transform the PLA into a modern military organization that relied for effectiveness more on technical expertise than on human-wave manpower and ideological drumming.

The PLA's Red Army traditions combined with the military modernization of the 1980s have resulted in an army that is caught between stages. The rank and file is still filled, as it always will be, with peasants who see more advantage in going into the Army than in staying at home. (A report in the Chinese press in April 1989 noted that as many as 85 percent of eighteen-year-olds were applying to join the PLA. This 85 percent is largely from rural areas, however, producing poorly educated recruits who do not measure up to the needs of a modern professional force.) But the officer corps is divided. At the senior level are Party stalwarts from semi-peasant backgrounds whose interests replicate those of the senior political leadership. The PLA's junior officers, on the other hand, tend to be from urban backgrounds, and better educated. So too does the technical staff attached to the PLA headquarters. These groups by and large backed the students' demands: as we noted in Chapter 2, the staff of the PLA General Logistics Department even marched in the very sizable parade in support of the hunger strikers on May 17.

Many Democracy Movement activists in May assumed that the professionalization of the PLA officer corps, combined with appeals to noble traditions of serving the people, would inhibit the Army from coming to Li Peng's defense. It was a plausible logic. Indeed, in Moscow two summers later, the world would see this logic yield results when officers of the Soviet Army decided not to back a coup against Soviet President Mikhail Gorbachev. The error in the Chinese assumption was to neglect the decisive power of the senior officer corps. The PLA is still run by men who owe their power and allegiance to Deng Xiaoping's faction within the Communist Party. Their allegiance is not abstract; most of them personally served in Deng's Second Field Army during the 1940s. It is in their interest to remain on his side, in spite of Deng's willingness to let the military's share of the national budget dwindle through the 1980s are well aware that, under a different dispensation, their careers l... __ finished. Thus, although the social composition of PLA officers straddles its peasant-revolution and professional-army stages, the older segment still prevails. But its days are numbered. As China moves further along in the present transition from the old revolutionary elite to the new technocracy, the Army will fall more and more into the hands of officers who identify with the reform process, and who opposed the use of martial law in 1989. As the years pass, the possibility grows that, come the next political crisis, the Army will intervene in favor of the reform faction. The spate of internal investigations, ideological campaigns, and leadership shuffles inside the PLA since June 1989 confirm that the Party's ruling faction is anxious to prevent this eventuality.

However we regard the opposition to martial law within the PLA, the issue dividing junior and senior officers is not democracy per se. It is the process of modernization: more specifically, how that process will affect the stability of the emerging technocratic elite. After all, Third World militaries have rarely shown themselves to be keen on opening the political process too wide. It is unrealistic, therefore, to expect the Army, shamed by its role in June 1989, to seek to regain its honor by fighting on the side of democratic change in the future. Junior officers may feel emotionally bound to the idea of democracy because of their experience in June 1989. But how they act the next time they are called upon to mediate domestic political shifts will depend far more on the particular political conjuncture in which they find themselves. Ideology may not play a determining role. Indeed, it is not impossible to imagine a more professional PLA a decade from now shooting people in the streets yet again—not to stall the transition from the old Communist order this time, but to ensure the successful implementation of market-oriented reforms. The interests of the military and the state may once again converge.

Those who participated in the Democracy Movement came to believe that they were taking history into their hands, that they were on the brink of committing China to profound change. A year after it was over, I spoke with one student about his sense of the Movement's significance. "This was the first time in history that the Chinese people gave their lives to transform the autocratic political system of centuries," he said, with pardonable exaggeration. "Since the Ming dynasty (1368–1644), many have died in civil wars, but these wars were only to change the emperor, not the system. June 4 is the first time people gave their lives to establish a democratic system. 1989 is the year China began to follow the course charted by the French Revolution, to die for freedom rather than for the emperor."

It was the youngest students, especially those in first-year university, but also in high school, who were the most emotional, the most radical, the most sanguine about the changes they thought they could bring about. For them, poised on a great ridge of hope, the Massacre was an unspeakable chasm into which to fall. A foreign scholar saw this fall in the teenage son of a colleague. "This kid was constantly on the phone with his friends the week after the Massacre. He was agitated, worked up, angry. His parents could not calm him down. The government had no credibility with the high school seniors in Beijing, the ones who would have gone to university the next fall." The older generation was just as disturbed. "Even conservative-minded intellectuals said that there had never been a government like this in the history of modern China," noted a graduate student. "Put together all the earlier incidents this century of governments suppressing students and it would be nothing compared with this."

How do we interpret the totality of these events—the hope infusing the Democracy Movement, the violence of the Beijing Massacre, the anger and utter disillusionment of the Chinese people? All stem from the nature of the challenge the Movement posed to the current dispensation of power in China. The students hoped that citizens' rights might be recognized as greater than the desires and privileges of a Party clique that dominates the state in the name of the people. This hope fortified them against the threat of force. They saw that threat as clear evidence of the Party's moral bankruptcy. It demonstrated that the Party's final commitment was to self-preservation rather than the interests of the people. The students insisted that, whatever contributions the Party had made in the past, these did not mean that the state "belonged" to the Party.

Similarly, under the students' inspiration, the citizens dared to believe that the city did not "belong" to the government; hence the Army had no authority to occupy the people's space in Tiananmen. Against the pure authority of the Party-as-state had arisen a sense of what has been called "civil society," the popular rights and social institutions that arose in eighteenth-century Europe in the realm between the family and the state. The concept of civil society expresses the idea that society—the realm of rational, noncoercive affiliations—has greater authority than the state. When the citizens of Beijing took the risk of resisting the Army to protect the students, they placed their own authority above the orders of the state. To resist was to act out the very ideas of rights and institutions that the Movement represented. Stopping the Army was itself a lesson in democracy. It was civil society in the making.

From his vantage point at the narrow summit of Party power, Deng Xiaoping was probably right to call in the Army. The Chinese Communist Party had to suppress the Movement if it wished to continue its monopoly of state power. Since 1949, the Party and the state have been the same entity. As the true representative of the people, the Party could claim to meld the state and the people into a unitary body. By separating the state from the people, the Democracy Movement called the Party's monopoly of power into question. Student rhetoric challenged this fusion of the people and the state by bringing forward concepts like democracy, human rights, and the independence of the press. All of these concepts erode the absolute authority of the Party-as-state and replace it, at least in theory, with a different authority, the people-as-state. The Party is seen no longer as the state representing all the people but simply as a political organization representing the particular interests of its members. For the Party to negotiate its rule on such terms meant capitulating its position of absolute authority, abandoning its role as state. It had no choice.

The Communist Party in China faces a conundrum cornering any ruling clique in the Third World that wishes to maintain itself in power. It must rule a nation in which no democratic consensus has ever been established. Nation-building occurred in China, as in many other places, as an elite reaction to international pressure in the twentieth century rather than through the grass-roots struggle to establish popular rights. The ruling clique thus has no basis of legitimation and must engage in constant work to ward off the suggestion that its rule is illegitimate. Through the claim that the Party represents the people, Communist ideology asserts that consensus has been reached, when in fact it has been imposed. The new authoritarianism popular with contemporary right-wing Asian regimes (and with some members of the Chinese Communist Party, including Zhao Ziyang) can be mobilized to achieve the same effect of making compliance appear to be part of the natural social order.

Although it is possible to fault Marxism-Leninism for furnishing an excuse for totalitarianism, ideology is not the real problem. Small oligarchies rule nations all over the Third World, under all manner of ideological persuasions. China is no different. We would better understand China's problems by focusing less on its government's ideology than on its predicament as a poor country in an international environment dominated by the developed capitalist world. In the end, the issue is not Communism. It is imperialism: the economic subordination of the underdeveloped world by the developed capitalist world.

Many analysts of the contemporary world reject the concept of imperialism as useful for describing the situation in which Third World countries find themselves today. I find it impossible to make sense of June 4, 1989, without it. The sheer desperation on both sides can only be accounted for by refocusing attention on the deep historical context of China's relationship with the world order emanating from the post-Renaissance European West. Since the sixteenth century, Europe has exploited the resources, labor, and ecological free-space of the rest of the globe to build up its wealth and power. China came under the influence of imperialism early in the nineteenth century, and its history since then has been the struggle to establish equal terms on which to engage in trade and diplomacy. That struggle has been economic, but it has also involved great political rethinking. Since the end of the nineteenth century, Chinese intellectuals have sought to create institutions that would enable China to respond more effectively to the social and economic problems of imposed backwardness.

Foreign pressure forced China to reorient itself away from a self-contained insularity to a world system whose core lay on the other side of the globe.

This reorientation has not been without resistance. The twentieth century opened in China with Beijing under siege by the Boxers. These poor men of north China turned to violence against the foreigners to protest their religious and economic intrusions. The Boxer Rebellion of 1900 generated no solution to the problem of China's relation to the outside world, except the reactive and ineffective one of driving foreign influences out of the country. (The foreign world was as censorious over the Boxer Rebellion as it was over the Beijing Massacre, seeing each as yet one more instance of China's failure to uphold international standards. The one English-language newspaper being published in China in 1900, the *North China Herald*, demanded of the imperial court that it undertake to ensure the safety of foreigners in China—much as foreign representatives eighty-nine years later were doing in the week after June 4. The imperialistic tone of the accounts Westerners later published of their experiences during the siege of Beijing makes me pause over some of the judgments in this book. Most foreign commentators liked to observe that the Boxers were hopelessly incompetent as fighters. Only Putnam Weale had the good sense to point out that "every siege in history has been like this—with everything incomplete and in disorder." Perhaps the PLA is no more hopeless than any other army; perhaps this is what always happens when men turn to arms.)

The xenophobia that fired the Boxers has continued as a dominant posture in state ideology during much of the People's Republic. A foreign enemy is always reassuring when domestic affairs are not going well. It is useful to a state that needs to distract its people from problems closer to home. Mao Zedong's campaign in the 1950s against American imperialism, as Deng Xiaoping's in the 1980s against spiritual pollution, pictured the West, in different degrees, as bent on destroying China. The intentions behind their uses of xenophobia differed, however. For Mao, it served as a means of resisting China's incorporation into the world capitalist market. Deng, on the other hand, accepts this incorporation. When he condemns Western influences as pernicious "bourgeois liberalization," it is to resist appeals to Western-based concepts and institutions that might challenge the Party's monopoly of the state.

The xenophobic theme resurfaced during the Democracy Movement, when the government attempted to mobilize public opinion against the activists by linking their ideas and actions to foreign sources of inspiration. The Goddess of Democracy was an easy target for this kind of misrepresentation. The statue's allusion to the Statue of Liberty was sufficient to discredit the aspirations of an entire generation. The theme is also played in a PLA account of a conversation between citizens and the troops they had immobilized on the evening of June 3. The citizens tell the soldiers that the government has used the media to dupe the soldiers, and that the soldiers would have a better understanding of what was going on had they been listening to the Voice of America. "We are Chinese," the soldiers are said to have replied, "and so we should listen to the voice of the Party Center. We think you should listen a little more to the voice of the Party Center, and not be fooled by foreigners." The argument about Chineseness tries to negate the Movement by making a xenophobic appeal. Good Chinese do not listen to foreign news because foreign news always serves foreign interests, and foreign interests are always inimical to China's interests. China's interests are best served by listening to only the Party Center. According to this logic, there can be no other truth.

The reforms of the 1980s sent a contrary message to the Chinese people. Foreign technologies and organizational theories have triumphed over the bootstrap approach of the Maoist years. And on their heels have come everything from impressionism to Coca-Cola. The opening to the outside world built into the reform policies cancels the power of xenophobic appeals.

The Party's other appeal against the demands voiced by the Democracy Movement is substantive rather than xenophobic. It argues that economic development in the Third World can be pursued only when political liberalization is kept to a minimum. This is precisely what Deng and his associates have endeavored to do: to develop a modern economy without adopting the political and legal institutions that developed in the West along with capitalism. The latter are luxuries that may come with time. For the present, it is the state, rather than private interest, that serves as the engine of growth. In this regard, they have acted in the latter half of the twentieth century as most Third World countries, regardless of ideological orientation, have acted. The political costs of developing an economy within a liberal environment are regarded as simply too great. Most Third World nations have had to revert to military force from time to time to keep their political arrangements in place. China is no different.

The Chinese government has argued openly for its right to limit the growth of individual freedoms. It has insisted that the perils of economic backwardness far outweigh the need for importing the individualist values of Western liberalism. Chinese government spokesmen express this idea when responding to international criticisms concerning China's violation of human rights. They pit the principle of respect for human rights against what they consider the greater principle of noninterference in the internal affairs of another country. They stress that the individual rights of citizens cannot take precedence over the collective rights of all to adequate food, health care, and education. As one apologist has put it, the tasks of post-colonial Third World governments are "to guard against the subversive and destructive activities of domestic and foreign enemy sources, to preserve social stability, and to relieve people's poverty and hunger." These are unfinished tasks, and until the collective right to live in stability is achieved, the luxury of individual freedoms cannot be granted. According to this way of thinking, the rights of the individual may not be furthered at the expense of the rights of all.

The roots of the nonrecognition of rights of the individual in Chinese law extend far back beyond the twentieth century. Imperial China developed complex codes and procedures to deal with legal matters, but almost entirely in relation to the enforcement of state regulations, not to the adjudication of private disputes—which is where civil law in Europe began. In imperial China, if an act was legally significant, then punishment followed. This is why the judicial administration under the emperors was called the Board of Punishments. In the West, on the other hand, if any act is legally significant, right of action follows. The old attitude of justice as punishment persists under Communist law, where legal procedures are largely limited to assessing penalties rather than assessing rights. As long as China has no independent judiciary, individual rights cannot be represented against the rights of the state. To rephrase the point made in the previous paragraph, the rights of all always prevail over the rights of the individual.

One can fairly argue that a nation whose hands are tied by poverty, external threat, and overpopulation must first attend to the right of collective survival before going on to address the right of individual action. Many within the Third World accept that a condition of fulfilling this task may be to abridge freedoms that are enshrined as rights in the wealthy Western democracies. But it requires an impossible leap of logic to invoke this appeal as a defense of the slaughter of civilians in the streets of the capital. Whatever predicament the Chinese government faces in handling its economic difficulties, that cannot justify the decision to send tens of thousands of armed soldiers against the people. There are always other choices. Other resources of persuasion, even of coercion, are available. There had to be other ways of reaching Tiananmen Square other than over the bodies of citizens. Indeed, a creative political solution might have strengthened the regime. Instead, force was used, thereby "robbing the Chinese government of its legality and legitimacy in the eyes of the people" as Wuerkaixi put it later.

Both disheartened Chinese and censorious spectators outside China should remember that neither China's leaders nor the students who sparked the mass movement asked the West to place China at the disadvantage it has faced for close to two centuries. It was not China's desire to be victimized by the capitalist world system. Nor has it been the wish of the Chinese people that the process of adapting to that system take the long and tortuous course it has, despite attempts of every ideological stripe to put things right.

On June 4, the leaders of China demonstrated that they lacked the political imagination to reestablish stability in any way except at a cost of thousands of lives. The dead remain unacknowledged, and the men who engineered their deaths unrepentant. This is so not because nothing has changed in this cycle of Cathay, not because the rulers living in the leadership compound in Zhongnanhai continue to think like the emperors who owned that pleasure park when Deng Xiaoping and Yang Shangkun were mere babies. Rather, Deng and his associates, like desperate modernizers elsewhere in the Third World, understand that history will judge their economic failures more harshly than their violations of human rights.

If this logic helps to explain what happened on June 4, it does not exonerate it. International standards of human rights—even if these standards were generated through Europe's struggle for political justice—still apply. Nor will any litany of appeals, either to China's present predicament or past traditions, ever explain away the surprise the world felt when China's leaders made this choice.

Two years to the day after the Massacre, I sat with a former Democracy Movement activist in front of the finely sculpted replica of the Goddess of Democracy unveiled outside the Student Centre at the University of British Columbia. Many people stopped that sunny morning, some curious to read the plaque at the base of the statue, others caught short by memories that are slipping into the dim reservoir of the past. One female student went to her knees and prayed for a moment in silence. As I talked with the exile about his experiences in the spring of 1989, I asked him what he thought of Deng Xiaoping now. "A criminal in the scales of history," he quipped in a mock voice of stern justice. Then he laughed. "Don't you see? It's not Deng who is at fault. That much power would corrupt anyone. Deng just happened to be the man in that position at that time." In the absence of genuine constitutional restraints, China's leaders are fated to be corrupted by the great economic and political power that lies in their hands. Deng has been no worse than the next autocrat.

Deng Xiaoping has banked on time to let the memory of the Massacre fade from his otherwise respected record as China's modernizing autocrat. Many are willing to conspire in the myth of his leadership and forgive him so long as he washes the blood off his hands. One of the rumors going around some months after the Massacre was that Deng's daughter advised him to "reverse the verdict on the Tiananmen Incident yourself while you are still alive rather than let others change it for you after you are dead." Surely Deng still remembers how completely he himself swept aside the legacies of Mao Zedong once the chairman was dead; surely he knows what will happen to his own reputation. Deng could make a reversal of verdict plausible by claiming he was duped by his subordinates, that conservative schemers manipulated his fear of being deposed to permit a decisive crackdown that would eliminate the power of the reform faction. Even Fang Lizhi offered Deng this out when he insisted to me two years later that "members of the Beijing Party Committee used the Movement to scare Deng into acting as they wanted him to. Only after the suppression did he realize that the students weren't that much of a threat. Perhaps he regrets the killing now."

The excuse that Deng was manipulated might be accepted for a man who looked as feeble as he did on television five days after the Massacre. But to admit to ignorance and manipulation may involve too great a sacrifice of Deng's precious and dwindling moral capital. In any other political system, and perhaps even in the Chinese, the admission would entail resignation. Emperors do not abdicate: they are the state incarnate. The man is not free to follow any other course.

Whether Deng Xiaoping decides to salvage his place in history is, of course, of little consequence to history itself. A reversal of verdict, even if undertaken for the wrong reason, might shift China slightly toward the political modernization that it badly needs. But in the end, even that shift may do little to steady China's crossing into the twenty-first century, for it is the world around China that sets the terms. We in that outside world may deplore the violence unleashed in the streets of Beijing. At the same time, however, we share the burden of having shaped a world in which the leaders of China could answer the challenge of youthful hope only with violence and despair.

Acknowledgments

This book grew out of research undertaken for the China Documentation Project. The idea of setting up a research project to investigate the Beijing Massacre emerged out of discussions among concerned China specialists in Toronto in the weeks following June 4, 1989. We felt that, this time, the specialists could not simply deplore events without doing anything about them. Our expertise obliged us to take up the task of finding out what really happened when the Army entered Beijing, and to consider what this might mean for China's future. This task became increasingly pressing in the face of the Chinese government's circulation of highly partial accounts.

The Project's focus has been to collect all types of documentation relating to the Beijing Massacre, and to the Democracy Movement more generally. Our main work was to track down and interview people who had been in China during this period, both foreigners and Chinese. Most of the debriefings were audiotaped on a confidential basis, with the understanding that access to them would be at the discretion of the project. My colleagues and I interviewed individuals in Canada, China, England, and the United States. Some whom we could not interview personally took the trouble to put their experiences down on paper and send them to the Project.

The original intention of the project was to produce a short report on the military aspect of the Massacre, which we felt was being ignored or misconstrued in the other reports then being produced. As information mounted up, however, it became apparent that a short report would not be adequate to the scale and complexity of the problem for which we were collecting material. A more extended study seemed to be required. The idea of writing a full-length book on the suppression took form in the course of discussions with project-member Lorraine Speiss, for whom the Massacre had an impact that was personal as well as professional. Accordingly, I decided to set aside my other research projects and take up the task. I trusted, perhaps naively, that my concern for the protection of human rights would compensate for the limits to my knowledge of Chinese military history.

Over the two years between conceiving and completing the manuscript, I have received the greatest possible support from Diana Lary. As director of the University of Toronto-York University Joint Centre for Asia Pacific Studies, she not only made the centre's facilities available to the project but gave generously of her time, her acumen, and, most important, her friendship to make sure this book got written.

Many colleagues and students read and criticized parts of the manuscript as it was being written. In particular I wish to thank Jeffrey Wasserstrom for his scholarly advice, even if I didn't always take it; we feel different parts of the elephant. I am also grateful to Richard King, who contributed valuable material to the project and helped me at the very end of writing to understand what I wanted to say.

Much of the inspiration for writing this sort of book came from my literary agent, Beverley Slopen. If it reads well at any point, the credit must go to Anne Montagnes and my fellow members of her Work Sessions in Creative Writing at the University of Toronto. Their critiques of the first and sixth chapters taught me to listen for my own voice. I will never ignore Gary Lewis's friendly warning not to write "like a dog shaking a dead doll," a redundancy even I had the wit to catch.

My deepest thanks, however, I reserve for those who were willing to unlock their painful memories of the Beijing Massacre. Most have chosen to remain anonymous. Those I can thank by name are Alex Day, Fang Lizhi, Steven Fredericks, Laura Fujino, John Gruetzner, Ruth Hayhoe, Gill Holland, Bengt Jörgen, Nicholas Jose, Tom Kennedy, Sheri Lecker, Pamela McCorduck, Perry Shearwood, Kenneth Smith, Roger Smith, Mike Sullivan, Wendy Tang, Stan Vittoz, Rudolf Wagner, Jan Wong, Wuerkaixi, and David Zweig. Without their dedication to truth and deep concern for China's fate, this book would not have been possible. It is their story, and I thank them for agreeing that it needed to be told. My most valuable informant on military matters must remain unnamed, but he knows who he is, and I am in his debt.

Financial support for research came from the University of Toronto, York University, and the History Department of the University of Toronto. A spontaneous contribution to the China Documentation Project from Helga Kasimoff is also gratefully acknowledged. The costs of completing the manuscript were covered by a grant-in-aid from the Social

Sciences and Humanities Research Committee of the University of Toronto.

Finally, on a personal note, it would never have occurred to me to undertake this work had it not been for the kind assistance of the staff at the Toronto office of Amnesty International. It is no exaggeration to say that, without Amnesty, the book would never have been written.

A Note on Sources

This book was initially conceived as an oral history of the Beijing Massacre. Most of the material is drawn from interviews with Chinese and foreign eyewitnesses who were in Beijing at various times between April 15 and June 9. Each is coded in the notes with the letters CDP (China Documentation Project) and a unique four-digit number.

Eyewitness testimony has the attraction of appearing to narrate events at first hand without the filter of hearsay and supposition. For the historian who seeks to reconstruct those events, however, it is not without its problems. Julian Barnes uses as the epigraph to his novel *Talking It Over* a Russian saying: "He lies like an eyewitness." When I read that line after struggling with eyewitness testimony for two years, trying to decide what information to accept and what to reject, the saying struck me with the force of revelation. Eyewitnesses do lie, persuasively, sometimes to protect themselves (especially if they are telling the same story to the Immigration and Refugee Board), more often because unknowingly they concede to faulty first impressions the status of truth. First impressions are so vivid that the eyewitness accepts without question the inferences she made about what she saw at the time she saw it. She clings to what made, and still makes, sense. What one recalls as an iron club, another remembers as an AK-47. Hindsight often deepens the confusion, as the eyewitness adjusts his assumptions to accommodate what others observed to him later on. Where one sees an angry street youth yelling insults, another remembers an undercover agent at work. There is in addition the problem of

civilian eyewitnesses making observations about military technical matters. It may not seem so important that an eyewitness tells me she saw a
tank when in fact she was seeing an armored personnel carrier. But such
errors can be critical to reconstructing events, since the two vehicles are
used for very different purposes and in very different settings—and hence
signify very different situations. My informants suffered from all these
shortcomings, each swearing by the validity of his own truth.

When I began to collect the material for this book, I did not anticipate
the difficulty of fitting together the flotsam of eyewitness impressions into
a composite image that was both complete and free from internal contradictions. Judgments regarding the admissibility of my informants' evidence had to be made at every turn. Sometimes I have been able to
corroborate eyewitness evidence by tallying it with other testimony; at
other times I have had to rely on my own larger sense of events to decide
on the reliability of what an individual has told me. Some of my informants will be dismayed to find that I have sometimes overridden their
observations with others, choosing against how they reported events to
me. I have done so only when the weight of other evidence has struck me
as overpoweringly contrary to their testimony. I do not doubt that I have
erred in some of the judgments I have made, though only fuller information about the events surrounding June 4 will convince me that my reconstruction is faulty. Until the people of Beijing feel free to bring forward the
trove of private diaries, letters, and photographs they have hidden away (as
CDP-1090 has assured me they have), these eyewitnesses speak with the
only voice of authority.

Occasionally, I have referred to rumors that eyewitnesses passed on to
me. I have done so only when they fit well with other testimony, and I
always specify their status as rumor rather than fact. Rumor is a valuable
but problematic source of information in China. With the official media so
tightly reined, unfit news, particularly regarding decision-making at the
highest levels of government, must travel by word of mouth. Rumor is a
reasonably efficient form of communication in China, having been the
principal source of information transmission there for forty years. (It is
also an indictable offense.) But the process of communicating through
rumor does introduce distortions and exaggerations that, once inserted
into a body of information, cannot easily be filtered out. Even so, rumors
almost always echo off certain walls of probability. The challenge lies in
reading the echo when one cannot see the wall.

I have supplemented eyewitness testimony with material from five
main sources: student leaflets and underground newssheets preserved at
the China Documentation Project and Columbia University, journalists'
articles, reports by Amnesty International and Asia Watch, the memoirs
of Democracy Movement activists, and Chinese government publications.
The latter two sources proved to be particularly difficult to use. Both

the Chinese government and the opposition have generated partisan litera-
tures designed more to dominate the public record than to set it straight.
Whereas the dissidents' accounts tend to exaggerate and emotionalize, the
government's accounts contain the more challenging, and more systematic, distortions. The problem the researcher faces is to evaluate not which
facts official writers choose to reveal but how they arrange them. Careful
editing frequently enables the official discourse to mislead without actually
telling any lies. The context is missing, and with it the real sense of the
scene described. It is like listening to someone describe a candle in a
hurricane without mentioning the wind.

Significant details often lie tucked inside the official narratives, nonetheless. The richest government source by far is *Jieyan yiri* [A day of
martial law]. This two-volume collection of soldiers' reminiscences produced by the Army's literary press was rushed into print four months after
the Massacre. Battle reminiscence is a familiar genre for Chinese readers.
A Day of Martial Law disturbs the genre by casting in the role of the enemy
not the reviled Nationalists or Japanese of the civil war period, but the
very people after whom the People's Liberation Army takes its name. It is
the citizenry—not the soldiers—who come across as righteous and brave in
these accounts. Comparing the PLA's maneuvers against Beijing with Red
Army attacks on enemy strongholds during the civil war (as is done in vol.
1, p. 85) only further confuses as to whose side the reader should be on.
What the book conveys to the unsympathetic reader—in other words,
most Chinese—is the commitment and integrity of the majority of the
demonstrators, and the duplicity and incompetence of the Army. Whether
this was the editors' intention is a matter of speculation.

A Chinese woman of my acquaintance remained aloof from the Democracy Movement while it was happening. She changed her mind after
reading *A Day of Martial Law*. Timid under normal circumstances, she
burst into my office one afternoon to announce her conversion to the
Movement. "I was so proud of how the people stood up against the
Army!" she exclaimed. Hers was not an isolated reaction, apparently:
the book was pulled from circulation shortly after it came out. "Never
before did I feel that the Chinese were a heroic people. I always thought
we were weak and easily pushed around. But now I know we are truly
great!"

This is not the book I would like to have written about the military
suppression of the Democracy Movement. That book would have ana-
lyzed the suppression in relation to the processes within the government,
the Party, and the Army that led to the decision to use military force.
Instead, I have written the book that could be written, working from the
oppositie direction: not from the decisions at the top down to their implementation (which was messy at best), but from partial and incomplete
records of what happened on the streets up to the thinking that prompted

the destruction. Some day, perhaps, researchers may gain access to more reliable sources—the internal memos of the Central Military Commission, for example, or the stenographic transcripts of the meetings of top Party leaders, or the log books of the group armies. In the meantime, we must live with the occupational bane of the social historian: the space between the phenomena of daily life he observes and the reality of power hidden behind.

Place Name Glossary

Beida ○ Beijing University in the northwestern suburb (15 km northwest)

Beishida ○ Beijing Normal University (8 km north)

Beitaipingzhuang ○ " intersection on the northern segment of the 3rd Ring Road (9 km north)

Changan Boulevard ○ "Boulevard of Eternal Peace," Beijing's main east-west thoroughfare; its formal name changes roughly every 2 kms.

Chongwenmen ○ "Gate of Literary Veneration," intersection at the east end of East Qianmen Street near the Railway Station (2 km southeast)

Dabeiyao ○ intersection where East Changan Boulevard crosses the 3nd Ring Road (6 km east)

Deshengmen ○ "Gate of Martial Victory," overpass on the northern segment of the 2nd Ring Road (6 km north)

Dongdan ○ intersection on East Changan Boulevard (2 km east)

Dongzhimen ○ "East Metropolitan Gate," overpass at the northeast corner of the 2nd Ring Road (7 km northeast)

Fengtai ○ satellite town southwest of Beijing (15 km southwest)

Note: Distance in kilometers and direction from Tiananmen Square appear in parentheses.

Forbidden City ○ the imperial palace at the north end of Tiananmen Square

Fuxingmen ○ "Gate of National Renaissance," overpass where West Changan Boulevard crosses the 2nd Ring Road (4 km west)

Gongzhufen ○ traffic circle where West Changan Boulevard intersects the 3rd Ring Road (9 km west)

Guangqumen ○ "Gate of Broad Greatness," intersection south of the southeast corner of the 2nd Ring Road (5 km southeast)

Haidian ○ northwestern suburb of Beijing, home of many leading universities

Hujialou ○ intersection on the eastern segment of the 3rd Ring Road (7 km northeast)

Jianguomen ○ "Gate of National Reconstruction," overpass where East Changan Boulevard crosses the 2nd Ring Road (4 km east)

Jianguomenwai ○ the main diplomatic compound just east of Jianguomen (4 km east)

Liubukou ○ intersection on West Changan Boulevard (1 km west)

Liuliqiao ○ overpass on the southwestern segment of the 3rd Ring Road (9 km southwest)

Madian ○ overpass on the northern segment of the 3rd Ring Road (8 km north)

Marco Polo Bridge ○ Lugouqiao, main bridge southwest of Fengtai (18 km southwest)

Muxidi ○ intersection on West Changan Boulevard between 2nd and 3rd Ring roads (6 km west); local pronounciation "Muxudi"

Nanyuan ○ airfield south of Beijing (11 km south)

Qianmen ○ "Front Gate," at the south end of the Square

Qijiayuan ○ diplomatic compound east of the Jianguomenwai compound (5 km east)

Qinghua ○ Qinghua University, in the northern suburb (16 km north)

Renda ○ China People's University, in the northwestern surburb (13 km northwest)

Shijingshan ○ city and district west of Beijing (25 km west)

Temple of Heaven ○ imperial park in the southeast quadrant of the old city (2 km southeast)

Tiananmen ○ "Gate of Heavenly Peace," at the south end of the Forbidden City overlooking the Square

Tongxian ○ city to the east of Beijing (20 km east)

Xidan ○ intersection on West Changan Boulevard (2 km west)

Xinhuamen ∘ "New China Gate," southern entrance to Zhongnanhai (1 km west)

Xinjiekou ∘ intersection north of Xidan inside the northwest corner of the 2nd Ring Road (6 km northwest)

Yongdingmen ∘ "Gate of Eternal Stability," main intersection at the southern end of the city (4 km south)

Zhongnanhai ∘ "Middle and South Lakes," an imperial park west of the Forbidden City, now the residence compound of the Chinese leadership (1 km west)

Notes

A	archival document held by the China Documentation Project, University of Toronto
ASA	Amnesty International external document on Asia
C:R	archival document held by the Starr Library, Columbia University
CBB	China Bulletin Board, an open computer information network through which Chinese students and supporters in North America shared information during and after the Democracy Movement
CDP	China Documentation Project confidential interview
FBIS	Foreign Broadcast Information Service, which translates and publishes a daily digest of news reports from foreign countries
JY	*Jieyan yiri* [A day of martial law] (Beijing: Jiefangjun Wenyi Chubanshe, 1989), 2 vols.

Chapter 1. Introduction

"Turmoil": Regarding the government's use of terminology to define its interpretation, see Wasserstrom and Perry, *Popular Protest and Political Culture*, especially Wasserstrom's Afterword.

"I never imagined that they would open fire": CDP-0618.

Soldiers inducted into the Army forty-three days before entering Beijing: JY 2:201.

More than 450 Tibetans killed in Lhasa on March 5–6: Jonathan Mirsky, *London Observer*, August 12, 1990.

Deng Xiaoping's comments, June 9: *Fourth Plenary Session*, pp. 12, 14.

China's cultural norms surrounding the use of violence: The question of the role of violence in Chinese culture has recently been opened for discussion with the publication of Lipman and Harrell, *Violence in China*.

Han Fei's reference to *The Art of War:* Watson, *Basic Writings*, "Han Fei Tzu," p. 110.

"Warfare is the Way of Deception": *Sun zi bingfa* [Master Sun's art of war], ch. 1. With hindsight, some participants in the Democracy Movement felt that the Movement's leaders would have done well to model their strategy on Master Sun in order to mount a more effective opposition: see Ding, *Minzhu yundong de Sun zi bingfa*.

"You speak of the value of plots and advantageous circumstances": translated in Watson, *Basic Writings*, "Hsün Tzu," p. 57.

"The ruler should not put his army in the field out of anger": *Sun zi bingfa*, ch. 12.

"Liable to rebound": Lao Zi, *Dao de jing*, ch. 30.

"Arms are implements of ill omen": *Dao de jing*, ch. 31.

"Turning back is how the Way moves": *Dao de jing*, ch. 40. Chinese government's response to Ceaucescu's execution: Associated Press, December 12, 1989; *South China Morning Post*, December 28, 1989; Asia Watch, *Punishment Season*, p. 70.

Chapter 2. Changing Fate

Order of the State Council imposing martial law: Oksenberg et al., *Beijing Spring, 1989*, pp. 315–16.

Martial Law Order No. 1: Simmie and Nixon, *Tiananmen Square*, p. 134.

"The students knew it was a terrific risk": CDP-0927

"Every popular democratic movement in the history of the world": *Xinwen daobao*, May 4 (A004).

Letters to Deng Xiaoping on behalf of Wei Jingsheng: Oksenberg et al., *Beijing Spring, 1989*, pp. 166–71.

"Hu Yaobang himself wasn't that important": CDP-0618.

"Young people couldn't help themselves": CDP-0716.

"Long live freedom!": FBIS-CHI-89-072, p. 26.

Changan Boulevard: Changan Boulevard changes name roughly every two kilometers. It becomes Jianguomennei Avenue, then Jianguomenwai Avenue to the east; Fuxingmennei Avenue, then Fuxingmenwai Avenue, then Fuxing Road to the west. For the sake of simplifying place names in this book, the entire length of this avenue within the 3rd Ring Road will be referred to simply as Changan Boulevard.

Demonstrators' names being taken at Beishida: CDP-0927.

Police take photographs on April 17: Fathers and Higgins, *Tiananmen*, p. 16.

"This was not a great movement for Western-style democracy": CDP-0927.

Scuffle at Xinhuamen on Thursday morning: FBIS-CHI-89-075, p. 19. This report also charges that the PAP used electric cattle prods.

New China News Agency report of April 27: FBIS-CHI-89-080, p. 11.

"He was beaten until his head was cut": FBIS-CHI-89-082, p. 45.

Leather shoes + leather belt: Simmie and Nixon, *Tiananmen Square*, p. 66.

"No-one lifted a hand!!!": A001.

"We miscalculated the number of students": FBIS-CHI-89-082, p. 47.

PAP playing with their belts: CDP-0907.

First-year students from the Beijing Police Academy put on active duty: CDP-0701.

The presence of soldiers in the Great Hall: CDP-0201.

Hong Kong report on military alert: *Cheng Ming*, May 1, 1989, in FBIS-CHI-89-083, pp. 28, 29; also June 1, in FBIS-CHI-89-103, p. 30.

"We saw that new recruits were mainly peasants": Shen Tong, quoted in Human Rights in China, *Children of the Dragon*, p. 48. For consistency, "peasants" has been changed from "farmers" in the original passage.

Canadian professor informed Beijing was "under martial law": CDP-0122.

Hong Kong Commercial Radio: FBIS-CHI-89-078, p. 19.

AFP: FBIS-CHI-89-079, p. 15. Twenty thousand is an exaggeration. The story was elaborated in the *Hongkong Standard* on Thursday: FBIS-CHI-89-080, p. 17.

PLA group armies: Their formation in 1985 is discussed in Lee, *China's Defence Modernization*, p. 25. Regarding the 38th Group Army, see p. 171.

Former soldier in the Beijing Military Region: CDP-0226.

Deng puts the Army on alert on April 25: Simmie and Nixon, *Tiananmen Square*, p. 37.

Stenographer's report of April 24 meeting: Kristof, "How the Hardliners Won," pp. 41, 66.

April 26th Editorial: FBIS-CHI-89-078, pp. 23–24.

"By any means": FBIS-CHI-89-079, p. 15.

PLA reform in the 1980s: see Lee, *China's Defence Modernization*, ch. 1; Rolph, "The PLA Army in 1987."

Neo-authoritarianism: The Chinese debate on this concept is surveyed in Oksenberg et al., *Beijing Spring, 1989*, pp. 123–49. For a fuller discussion, see Ma, "The Rise and Fall of Neo-Authoritarianism."

"It was widely rumored that the head of the PSB": CDP-0701.

Fifteen trucks full of police reinforcements: AFP stated on the basis of student reports that the trucks of uniformed men parked in the Square earlier that day were from the 38th Army, but this cannot be confirmed; FBIS-CHI-89-082, p. 8.

"A brilliant decision": Wakeman, "The June Fourth Movement," p. 59.

"The procession had already turned up the 2nd Ring Road": CDP-1010.

"The soldiers looked disoriented": CDP-0717.

Army leaves canceled: CDP-0718.

New China News Agency account of April 27 march: FBIS-CHI-89-080, p. 18.

"Friends of one of the students in intensive care": CDP-0717.

Hongkong Standard report of missing students: FBIS-CHI-89-082, p. 56.

"Blood will be shed": CDP-0122.

"We were nearing the seventieth anniversary": CDP-1030.

"I have seen the students on the street": CDP-0504.

Zhao Ziyang's May 3 speech: FBIS-CHI-89-084, pp. 16–19.

"There was no show of force by the Army": Shapiro, "Letter from Beijing," p. 79.

"Absolutely no hostility": CDP-0122.

Yuan Mu's comment on May 12: FBIS-CHI-89-091, p. 21.

"The idea of a hunger strike": Zhang, "Democracy versus Freedom," p. 8. For an examination of the hunger strike as an innovative addition to the protest repertoire of Chinese students, see Esherick and Wasserstrom, "Acting Out Democracy," p. 841.

"The ones who were fasting sat themselves down": CDP-1021.

Collapse of hunger strikers: FBIS-CHI-89-093, p. 44.

Statistics regarding hospital treatment for hunger strikers: Zhonggong Beijing shiwei bangongting, *1989' Beijing zhizhi dongluan*, p. 108.

"The key was those seven days": CDP-1030.

Qinghua students appeal for PSB help: FBIS-CHI-89-095, p. 55.

Beijing PBS banner denouncing police files: *New York Times*, May 19.

Rumors of plainclothes policemen: CDP-0730.

PLA General Logistics Department: FBIS-CHI-89-094, p. 57. At various times, other Army personnel participated in the demonstrations, including students at the PLA Armored Engineering College, Army ambulance and medical personel, staff from a PLA hospital, and members of the General Staff, as well as demobilized soldiers.

"The officers and men": C:R-0.

Commander Xu Jingxian: C:R-9. It is alleged that General Xu was confronted in hospital by Zhou Yibing and Liu Zhenhua, Commander and Chief Political Commissar respectively of the Beijing Military Region.

PLA officers' open letter to the CMC: FBIS-CHI-89-095, p. 38.

Gorbachev's comments on the students' activities at his press conference: FBIS-CHI-89-095, p. 19.

"When we saw Li Peng talking": CDP-1030.

"I have the Army behind me": Fathers and Higgins, *Tiananmen*, p. 70; CDP-0716.

A former Army officer regarding Li Peng's daughter: CDP-0718.

"A lion just now waking up": CDP-0518.

"The Communist Party, like a strong tree": CDP-0518.

Zhao Ziyang's comments on the Square on May 19: FBIS-CHI-89-096, pp. 13–14.

Zhao Ziyang at the Politburo meeting on Wednesday, May 17: Chen Yizi, quoted in Human Rights in China, *Children of the Dragon*, p. 79.

New China News Agency report on May 19: FBIS-CHI-89-096, p. 26.

Hong Kong Commercial Radio: FBIS-CHI-89-096, p. 10.

"Things had quieted down by Friday": CDP-1021.

Martial law plans leaked by an aide to Zhao Ziyang: as reported in the June 30 speech by Mayor Chen Xitong, in Zongzhengzhibu xuanchuanbu, *Hanwei shuhuizhuyi gongheguo*, p. 519.

"He was absolutely furious": CDP-1021.

"When we heard about martial law on May 19": CDP-1020.

"Li Peng will never get away with it!": CDP-0701.

"Unconstitutional Regulations Have No Legal Force": A039.

Rumor at People's University that Zhao had resigned: CDP-0701.

Four minibuses stopped near No. 4 Hospital: JY 2:2. Li Peng's and Yang Shangkun's May 19 speeches: FBIS-CHI-89-097, pp. 9–12.

Yang Shangkun: FBIS-CHI-89-097, p. 19. Yang was involved in placing the 38th Group Army in a state of readiness in January 1987 during the previous student movement in Beijing; on that occasion the Army was never brought into the capital. In the CMC reorganization in November 1989, when Deng gave up the CMC chairmanship, Yang kept his position as the first vice-chairman. Yang Baibing was promoted to become the CMC's secretary-general. With the politically powerless Party Secretary-General Jiang Zemin in the CMC chairmanship, the Yangs continue to have overwhelming control of the CMC. There may be another Yang family connection in the military; it was widely reported during the Democracy Movement that Chi Haotian, chief of the PLA General Staff since November 1987, is Yang Shangkun's son-in-law.

"If a group of political commissars had not insisted on their political stand": *South China Morning Post*, December 28, 1989.

CMC announcement on May 23 that Yang had the right to transfer troops: *Cheng Ming*, no. 140 (June 1, 1989), in FBIS-CHI-89-103, p. 30.

Reports of Deng's trips to Wuhan: FBIS-CHI-89-097, pp. 4-5; FBIS-CHI-89-098, p. 16.

Chapter 3. No Place Left Unguarded

Team of troubleshooters sent toward Gongzhufen at 8:40 P.M.: JY 2:1.

Scuffle at Gongzhufen on Saturday morning: reported by Kyodo, FBIS-CHI-89-097, p. 43.

PSB and PAP attack demonstrators at Liuliqiao: FBIS-CHI-89-097, pp. 40–41, 82.

Two dozen students injured: Six of these students were from the Second Foreign Languages Institute; one was hospitalized (A031).

Soldiers from Shanxi province (near Taiyuan): CDP-0718.

Pierre Hurel: *Paris Match*, June 1, p. 59.

Contingent from the south: FBIS-CHI-89-097, p. 87.

"The young soldiers standing in the back grinning sheepishly": CDP-0622.

Qianmen barricade: FBIS-CHI-89-097, p. 43.

Armored convoy from the east: *New York Times*, May 21, reported this as a tank convoy, but this is unlikely.

Dabeiyao blockade at 3:00 A.M.: Munier, *Voyage au Printemps*, p. 32.

"Who is causing this unhappiness for the people?": *Paris Match*, June 1, p. 58.

Identification of trucks as antiaircraft resupply vehicles: CDP-1022.

Blockades as far east as Tongxian: JY 1:1.

Blockade at Hujialou intersection: A065; FBIS-CHI-89-097, pp. 43, 44; FBIS-CHI-89-099, p. 32; JY 1:282.

"They had come in from Zhangjiakou": CDP-0705.

Report of sixty-five APCs and seventeen tanks: FBIS-CHI-89-097, p. 54.

Beijing Party Committee urgent order at 6:00 A.M.: Zhonggong Beijing shiwei bangongting, *1989' Beijing zhizhi dongluan*, p. 96.

Truck column blocked toward the Fragrant Hills: JY 1:26.

"We have towels for tear gas": *New York Times*, May 21. A foreign observer at People's University recalls that the students also had towels for protection against tear gas that local merchants donated to the Movement (CDP-1021).

Students on the Square wait for the Army to arrive: CDP-1062.

Robin Munro: quoted in Human Rights in China, *Children of the Dragon*, p. 102.

"I have good news for you all": FBIS-CHI-89-097, p. 25.

Student leaflet distributed to troops on Saturday: C:R-5.

"Five helicopters flying in formation": CDP-1022.

"PAP and PLA soldiers have the right": Human Rights in China, *Children of the Dragon*, p. 103.

"The leaflets warned people that they should get off the Square": CDP-1031.

"The students there were certain": CDP-0201.

"It was fantastic to see": Munier, *Voyage au Printemps*, pp. 33, 36.

Squads of students spreading news around the city: CDP-0622.

"We'll never hurt the students": CDP-0907.

"At that time, the soldiers were quite polite": CDP-1020.

Soldiers carrying AK-47s on trucks: CDP-0805.

Speculation regarding whether AK-47s were loaded: CDP-0718.

Four thousand people stop ambulance at Xinjiekou: JY 1:5.

Regiment blocked at Tongxian: JY 1:4.

Rumor of parachute assault on the Square: Munier, *Voyage au Printemps*, p. 41.

Students organize blockade of Temple of Heaven: Zhonggong Beijing shiwei bangongting, *1989' Beijing zhizhi dongluan*, p. 101.

Subway shut down: FBIS-CHI-89-097, p. 54; CDP-1009. On May 26, a subway spokeswoman denied that the subway had been used to move troops the previous weekend and suggested that the timing of the closing with martial law was pure coincidence (FBIS-CHI-89-101, p. 34).

Public buses driven into Qianmen subway exits: *Testimonial to History*, p. 28.

Troops moved through the tunnel network: FBIS-CHI-89-097, p. 54; CDP-1022.

Troops in the Beijing Railway Station: FBIS-CHI-89-097, pp. 47, 75.

Troops at the Shahe Railway Station: CDP-0705.

Troop train blocked at Changping Station: JY 2:10.

Trains blocked at Qianan and Jinzhou stations: Jy 2:6–7.

"Within Beijing, the populace exercises": Munier, *Voyage au Printemps*, pp. 41, 42, 43.

"The concrete road dividers made Ss": CDP-0201.

Two thousand people block oil tankers at the entrance to Nanyuan Airfield: Zhonggong Beijing shiwei bangongting, *1989' Beijing zhizhi dongluan*, p. 101.

Soldiers at the *People's Daily*. FBIS-CHI-89-097, p. 56; JY 1:1.

Soldiers infiltrated from Shunyi: JY 1:11–13.

"A Notice to the People of Beijing from the Martial Law Command of the PLA": A025.

List of slogans from the Martial Law Command: A027.

List of slogans from Beida: A050.

"A small group of bureaucrats": A028.

"The students are guilty of no sin": A051.

Open letter from an "old soldier": A040.

Soldiers ordered in without much knowledge: FBIS-CHI-89-097, p. 82; FBIS-CHI-89-098, p. 20.

Student leaflet condemning the April 26th Editorial: C:R-8.

"I don't know": CDP-0718.

"I will absolutely not use my weapon": FBIS-CHI-89-097, p. 82.

Interview with Commissar Liu Zhijun: New China News Agency Domestic Service, May 24, cited in FBIS-CHI-89-100, p. 28.

Open letter of seven retired generals: FBIS-CHI-89-097, pp. 23–24, 28; FBIS-CHI-89-099, pp. 18–19; *New York Times*, May 23. The signatories were Zhang Aiping, Xiao Ke, Yang Dezhi, Ye Fei, Chen Zaidao, Song Shulun, and Li Jukui. Zhang Aiping and Yang Dezhi were active proponents of military modernization through the 1980s; Yang had called for a "democratic" mood within the PLA (Lee, *China's Defence Modernization*, pp. 509, 226). Yang Dezhi died in the autumn of 1990.

Nie Rongzhen and Xu Xiangqian: FBIS-CHI-89-097, pp. 51–52.

State television discounts rumors of attack: FBIS-CHI-89-097, p. 30.

"All the thoroughfares were completely blocked off": CDP-1020. Barricades at intersections are mentioned in FBIS-CHI-89-098, p. 57.

Deng Pufang as "crown prince" (*taizi*): A002.

Poster warns Deng Pufang that he would lose the other leg: CDP-0608.

Wuerkaixi resigns: FBIS-CHI-89-098, p. 28.

"That night we feared": CDP-0201.

"When I got the report": CDP-0207.

"In Western countries": CDP-1020. The same belief was mentioned by CDP-0701.

Beijing Television interview: FBIS-CHI-89-097, pp. 57–58.

"The Army column just south of Liuliqiao": CDP-0201.

Fighting near Liuliqiao: FBIS-CHI-89-098, pp. 41, 43.

Twenty-nine officers and soldiers hospitalized: FBIS-CHI-89-101, p. 31. No. 304 Hospital is mentioned again in the Chinese media two days later: FBIS-CHI-89-103, p. 59.

Junior officer dies on May 23: FBIS-CHI-89-100, p. 25.

Naval cadets in Tuesday demonstration: FBIS-CHI-89-099, p. 29.

"Cancel Military Rule": photograph in *New York Times*, May 23.

Municipal government spokesman: FBIS-CHI-89-097, pp. 30–31.

The Art of War: Sun zi bingfa, ch. 11.

Troop truck from Pingshan County: Zhongyang Beijing shiwei xuanchuanbu *Pingxi baoluan*, p. 40.

"We found that they had been on the road for several days": CDP-0705.

"They had been in their trucks for five days": CDP-0201.

Vice-commander receives order to fly to Beijing at 7:30 A.M.: JY 1:185.

Students of Hebei University in Baoding block the 27th Group Army: Jochnowitz, "True Power to the People," p. 22.

Information regarding unit identification has been taken in part from FBIS-CHI-097, pp. 6, 24, 54; FBIS-CHI-89-098, p. 41; FBIS-CHI-89-099, p. 19; *New York Times*, February 15, 1990.

Shanghai City Service report on May 25; FBIS-CHI89-100, p. 23.

"Do you have anything to say?": C:R-9.

"The entire officer corps of the 38th Group Army": C:R-5. A similar statement appears in another leaflet, C:R-7.

Views on the 38th Army: FBIS-CHI-89-097, pp. 17, 24. 41.

Foreign scholar in Changchun: CDP-0122.

Hongkong Standard: FBIS-CHI-89-097, p. 5. The report declared that the commander of the Lanzhou Military Region also sent Deng a letter.

Guangzhou Military Region: FBIS-CHI-89-099, p. 68. One of the most important military districts within this region, the Hubei Military District in Wuhan, held a similar meeting on the same day: FBIS-CHI-89-101, p. 53.

Nanjing Military Region: The support of the Jiangxi Military District was announced on May 22: FBIS-CHI-89-102, p. 38. Military commands elsewhere in the country issued notices of support for the declaration of martial law in Beijing in the first few days after May 20, some noticeably more quickly than others. The Hong Kong magazine *Cheng Ming* later stated that the 12th Group Army from the Nanjing Military Region was part of the initial martial law force (FBIS-CHI-89-103, p. 31), but I have no corroboration for that statement. On the political character of the Nanjing and Guangzhou military regions, see Lee, *China's Defence Modernization*, p. 241.

"Several days of earnest study": FBIS-CHI-89-100, p. 40.

"Soldiers were being moved in from other regions": CDP-0701. The Hong Kong newspaper *Ming Pao* carried a similar speculation: FBIS-CHI-89-098, p. 19.

"A handful of conspirators": *New York Times*, May 28.

Letter from an infantry major: A057.

"When martial law was declared": CDP-1061.

"The Movement continued to grow daily": CDP-1020.

"Every day for the next week": CDP-1024.

"Strictly observe discipline": from a circular issued on May 20 by the Qinghai Military District in the Lanzhou Military Region: FBIS-CHI-89-100, p. 60. Two days later, the Xinjiang Military District issued a similar circular to its security forces: FBIS-CHI-89-100, p. 40.

"People were out persuading": CDP-0717.

"I think our soldiers have behaved well": FBIS-CHI-89-098, p. 40.

"Of course we weren't afraid": CDP-0705.

"That which is not of the Way": Lao Zi, *Dao de jing*, ch. 30.

Chapter 4. Waiting for the Moon

Two encirclements: reported in *Wen wei po*, FBIS-CHI-89-102, p. 53.

Troops encamped in the western suburbs: *China Daily*, quoted in FBIS-CHI-89-100, p. 17.

Maximum alert: reported in *Ming pao*, FBIS-CHI-89-101, p. 27.

"It's a full moon": Holland, "Beijing Spring," p. 37.

Units newly arrived in the Beijing area: The 39th from the Shenyang Military Region arrived by train to join the 16th, which had been too late to take part in the first assault. The 40th, also from the Shenyang Military Region, must have arrived by this time, though it is not identified among the troops in the capital until Sunday afternoon (Yang Jianli, quoted in Human Rights in China, *Children of the Dragon*, p. 183). The 54th and 67th from the Jinan Military Region were joined by the 20th and 26th, with the result that every group army in that neighboring military region was represented in the martial law force. Jane's report, pp. 19–23, asserts that units from the 1st Group Army (the Nanjing Military Region), the 14th (the Chengdu Military Region), the 21st (the Lanzhou Military Region), and the 64th (the Shenyang Military Region) were also brought in, but I have found no confirmation that these units were involved. The Jane's report was produced hurriedly and is unreliable.

Seven hundred soldiers at Shahe Station: FBIS-CHI-89-100, p. 17.

Supplies being flown into Nanyuan Airfield: FBIS-CHI-89-100, p. 20.

Troops airlifted on Thursday night: reported by Stuart Pallister for Hong Kong Asia Television, FBIS-CHI-89-100, p. 24.

No airplane tickets to be issued in Guangzhou for six days: FBIS-CHI-89-100, pp. 32–33.

Higher estimates of troop numbers: Landsberger, "Chronology," p. 174, gives 300,000.

Tanks detected forty kilometers east of the city: CDP-1022.

Enlarged meeting of the CMC: Landsberger, "Chronology," p. 176.

People's Daily, May 25: FBIS-CHI-89-100, pp. 16–17. The support the Navy gave to martial law on the front of the *People's Daily* tallies with its known loyalty to Deng Xiaoping. It proved later to be even more tangible: one of the videos produced by the government to present its version of the suppression shows naval marines in Beijing in the days after the Massacre.

New China News Agency on the Beijing Military Region: FBIS-CHI-89-101, p. 25.

Poster calling for Li Peng to step down pasted up at the headquarters of the Beijing Military Region: *Cheng Ming*, no. 140 (June 1, 1989), in FBIS-CHI-89-103, p. 28.

Local donations to military units: FBIS-CHI-89-103, pp. 59–61.

Shahe county government billets soldiers: FBIS-CHI-89-103, p. 60.

May 27 *News Herald* (Xinwen daobao): A072, p. 2.

The NPC and martial law: Leading reformist intellectuals Yan Jiaqi and Bao Zunxin published an article on the constitutionality of martial law on May 26 in Hong Kong's *Ming pao*, translated in FBIS-CHI-89-101, pp. 11–12.

Fifty-seven members of the NPC Standing Committee: FBIS-CHI-89-100, p. 22. The signatories included Hu Jiwei and such leading intellectuals as writer Wang Meng, political economist Wu Dakun, and veteran historians Zhou Gucheng and Liu Danian. To judge from a slip he made regarding the imposition of martial law in a public address on Friday, May 26, Li Peng himself was, if not confused about the constitutional regulations on martial law, at least sensitive to the criticism that he was proceeding unconstitutionally. The text of this address is translated in FBIS-CHI-89-101, p. 10.

Students' open letter to Wan Li: A064.

Students wait for Wan Li at the Beijing Airport: FBIS-CHI-89-100, p. 21.

Confidential source inside the Beijing Airport: CDP-0730.

Wan Li's statement: FBIS-CHI-89-102, pp. 30–31.

Radio Beijing's Monday evening broadcast: CBB, News Report from the University of Regina, May 29, 22:57:42 CST.

The municipal government orders the formation of "inspection teams": A062.

Open letter from the students to the workers: A063. All-China Federation of Trade Unions: Landsberger, "Chronology," p. 172; *People's Daily*, May 23, in FBIS-CHI-89-100, p. 31.

Activists leaflet factories: CDP-0604.

Strike activity at Capital Iron and Steel: CDP-0730.

Workers charged with "inciting" steelworkers: the trial of Lin Qiang and Wang Liqiang on June 30, 1989, was reported on July 6 in *Beijing wanbao;* see Asia Watch, *Repression in China*, p. 50.

Central leaders visit Capital Iron and Steel: CDP-0604. Several sources assert that Li Peng was among this group: CBB, Beijing Boston Hotline News Update, May 22, 18:38:39 GMT; Landsberger, "Chronology," p. 173.

Dare-to-Die Squads: The Beijing Citizens Dare-to-Die Squad announced its formation on May 26 (A069). Another, the Beijing Workers Dare-to-Die Squad, was in operation at least as early (A071).

"They were young": CDP-0730.

"Worst for both sides": Lu, "Beijing Diary," p. 12.

The making of the Goddess of Democracy statue: CDP-0806.

Police make no appearance at the Goddess's unveiling: Lu, "Beijing Diary," p. 8.

Steelworkers organized to knock down the Goddess of Democracy: reported in *Ming pao*, FBIS-CHI-89-104, p. 21.

Students' Federation urges out-of-town students to return home: A077 (May 29).

"Most Beijing students, including the Students' Federation": CDP-0716.

Group army headquarters approaches student leaders to end their occupation of the Square: JY 1:84.

Chinese military cameramen ordered to prepare on May 26: CDP-0907.

Order on May 26 to infiltrate to the Military Museum: JY 2:27–29.

"They were in uniform but not armed": CDP-1215.

Troops at key locations in the city: reported by New China News Agency Domestic Service, FBIS-CHI-89-104, p. 26.

"The staff was aware of their presence": CDP-0801.

Officers' meeting at Nanyuan Airfield on May 31: JY 1:187.

Occupation plans settled by May 31: The Wednesday edition of the *South China Morning Post*, citing "sources close to the military," suggested that the Army "may be entering the capital as early as the end of the week," though it said that the exact timing would depend "on the official perception of the level of resistance put up by students and citizens."

8:00 A.M. tourist train canceled on June 1: CDP-1009.

Word of troop movements passed among students at noon on June 1 (Lu, "Beijing Diary," p. 11).

"There were constant announcements": CDP-1009

The national radio news hookup, May 29: A079 (May 29).

"The students are the only ones speaking the truth": CDP-1090.

Government-sponsored demonstration at Daxing: CDP-0907.

Students' counterdemonstration in Changping: FBIS-CHI-89-104, p. 19.

"I didn't take part in any of the students' organized activities": CDP-0618.

Workers' resistance to participating in pro-martial-law demonstrations: A085 (May 31, June 1), A091 (June 2). The attempt to pay urban workers to join demonstrations on Wednesday was also mentioned by CDP-0806.

Hunger strikers' manifesto: see Han, *Cries for Democracy*, pp. 349–54. Zhou Duo, arrested on July 7, 1989, was released on May 9, 1990, as part of a general release of 211 prisoners.

"When word spread that Hou Dejian was leading singsongs": CDP-0801

Car without license plates stopped: "Tiananmen guangchang baowei bu-zhang," p. 24.

Divisional commanders driven to the Great Hall after midnight: JY 1:64.

Ten thousand soldiers in plainclothes enter the Great Hall on Friday: JY 1:56. This division included an artillery brigade.

Operation F75: JY 1:60–63.

Soldiers moving by foot from Fengtai on Friday: JY 1:80.

The student command vehicle usually at Dongdaqiao not there on Friday night: CDP-0925.

"The trucks were empty": CDP-1031.

Troop trucks near the Beijing Zoo: CDP-0917.

Identification of the soldiers as 24th Group Army: CDP-0806.

The Mitsubishi jeep accident: CDP-1022; A093 (June 3).

"The bikes and bodies were all twisted up": CDP-0917.

PSB cruiser spirits soldiers away: A101 (June 3).

"We were safe before the Army entered the city": A093 (June 3).

"I remember this clearly": CDP-0215.

"An important signal that the Army might be about to move": CDP-0716.

"Don't stay here. Something nasty is going to happen": CDP-0917.

Two thousand soldiers blocked along Xuanwumen Avenue: JY 1:80.

"Groups of people ran along beside them": CDP-0917.

"Don't worry": CDP-0207.

"The soldiers were running in short-sleeve white shirts": CDP-1022.

Speculation that the soldiers had jogged from Tongxian: A093 (June 3).

Two sets of barricades behind the crowds at Wangfujing: CDP-0917.

"When people yelled at them to sit down": CDP-0907.

"The crowd would not let them leave": A093 (June 3).

"They were running away from the direction of the Square": CDP-0907.

Others struck that the soldiers seemed to be without commanding officers: CDP-0801.

"The ratio of officers to soldiers was very small": CDP-1022.

"A man came running up to me": CDP-0801.

Citizens come out in night dress: CDP-0801, CDP-1022.

Soldiers surrounded in the south end of the city: Zhongyang Beijing shiwei xuanchuanbu, *Pingxi baoluan*, pp. 84–86. This incident occurred on Dongjing Road: The 24 wounded were removed to No. 302 Military Hospital in disguise Saturday evening.

"First we went to the Square": CDP-0805.

Infiltration of troops by bus after 1:00 A.M.: JY 1:113, 121–23.

Dozens of buses carrying soldiers stopped: A093 (June 3). The same source declares that eight peasants were run down by a convoy northwest of the city about

4:00 A.M., and another four at 7:50 A.M., but I have found no independent confirmation of this claim.

Bus full of soldiers stopped north of the Forbidden City: CDP-0917.

Driver sneaks away on washroom break: JY 1:116.

"The soldiers were not in uniform": Lu, "Beijing Diary," p. 14.

Driver sticks his officer's pistol in his underpants: JY 1:114.

"Some of the students and workers": CDP-0929.

"On closer examination, you could see": CDP-0717.

Infiltration of soldiers in single trucks: A097 (June 3).

Lost soldiers gathered up by plainclothes agents: JY 2:393.

Thirty soldiers sent to recover weapons: JY 1:81–82.

Students hand over weapons to PSB to be deposited at the Navy Hospital: JY 1:83, 135.

One soldier admitted to the Beijing Municipal Emergency Center: Zhongyang Beijing shiwei xuanchuanbu, *Pingxi baoluan*, p. 65.

Teargassing at Liubukou: A096 (June 3).

Soldiers misjudge the wind and the tear gas blows back in their faces: CDP-0603.

Two elderly residents count tear-gas firings: CDP-0917.

Forty protesters injured: Duke, *The Iron House*, p. 99.

At least twenty admitted to No. 2 Hospital: A101 (*Xinwen daobao*, June 3), p. 1.

"Metak 38m MN-05": *Time* Magazine Editors, *Massacre in Beijing*, p. 24.

"I saw a guy who had been hit in the face": CDP-1062.

"People were throwing rocks at the uniformed men": CDP-0801.

Tear-gas attack at 4:40: CDP-0801.

"We saw a man about thirty or so": CDP-1030. A Beida handbill (A096) states that this girl was blinded by the tear gas, but the statement cannot be confirmed. The blinding was probably temporary. A098 also reports on the Beida student march.

"The street was a sea of humanity": CDP-1030.

"The street was mobbed": CDP-0801.

"A space had been cleared between the steps": CDP-1032.

"Dogs! Creeps!": CDP-0907.

"A deliberate provocation": Duke, *The Iron House*, p. 96.

"It makes no sense to send in a truck": CDP-1062.

"The Army photographers were out": CDP-0207.

No activity at the PLA camp at Mentougou: CDP-1022.

Chapter 5. Spilling Blood

The government loudspeaker system is cut: CDP-1062; also reported by CDP-0603.

Reactions to the emergency statement at 6:00 P.M.: CDP-0716.

Warning of military operation telephoned to student headquarters at 4:00 P.M.:

"Beijing Spring 1989," p. 121; Chinese version reprinted in Han Shanbi, *Lishi de chuangshang*, p. 157.

Meeting of heads of group armies at 4:00 P.M.: JY 1:84.

Decision to use one of the PAP riot squads made only at 5:30: JY 1:92.

Units outside the city's perimeter start to move about 3:00 P.M.: A unit left its camp in southeastern Daxing county at 3:10 (JY 2:165). A truck column northeast of the city received the order to go into the city at 3:30 (JY 1:241). A regiment based at Shahe Airfield received the order at 3:35 (JY 1:224). A brigade of fifty-five trucks to the northeast received the order at 4:00 (JY 2:107). A company posted in Nanyuan Airfield received the order at 4:15 (JY 2:201). A unit of forty-one troop trucks and over eight hundred soldiers at Shahe received the order at 4:45 (JY 1:233).

Order to begin offensive issued at 5:00 P.M.: JY 1:84, 188, 351.

Truck column stopped at the northeast corner of the 4th Ring Road: JY 1:241.

Column of fifty-five trucks stopped outside 4th Ring Road: JY 2:107–109.

Regiment blocked on Airport Road: JY 1:225.

"Using your opponent's spear to attack his shield": JY 1:234.

"Eighteen trucks full of soldiers": CDP-0907.

Three thousand troops observed leaving Shahe Airfield: CDP-1022.

"We saw upwards of a hundred open-backed troop trucks": CDP-0928.

"Citizens of Beijing: We have come to protect order in the capital": JY 1:100.

"You must talk to the commander": CDP-0717.

Thirty-five trucks stopped at Jianguomen at 6:15: CDP-0929, CDP-1031.

Troops at Jianguomen said to be from Shenyang: CDP-0801.

Column sets out from Tongxian at 5:50: JY 1:157–58.

Roads out of Tongxian blocked by 7:00: JY 2:302. The account in which this is mentioned tells of a medical unit consisting of one ambulance and five trucks that was ordered at 6:30 to follow the group army to which it was attached. When the unit reached Beiyuan in Tongxian at 7:00, it was blocked by several thousand people.

Head of column reaches Jianguomen at 7:10: JY 2:447.

Column coming from the southeast blocked along the 3rd Ring Road: JY 1:256–57. The two intersections south of Dabeiyao are Shuangjing and Jinsong.

Column coming from Sanjianfang Airfield: CDP-1022; JY 2:138.

"One truck got away": CDP-1032.

"We got there just as they started to jump off": CDP-1031.

"Anyway, we have the soldiers surrounded": CDP-1022.

Video footage of the workers' truck: "Piaoyang, gongheguo de qizhi."

Workers distributing weapons on the Square included in official account: Zhonggong Beijing shiwei bangongting, *1989' Beijing zhizhi dongluan*, p. 126.

"Barricades in the street the whole way down": CDP-0801.

Guns in the backs of trucks at Jianguomen: CDP-1032.

"What do you think they've got guns for?": CDP-0718.

"We had just finished dinner": CDP-0201.

Trucks travel by country roads shorn of military plates: JY 1:154–55.

Trucks burned near People's University: CDP-0917.

Orders to move received by units in the south at 4:15 and 5:00: JY 2:201, 1:188.

Company of 122 soldiers: JY 2:201–202.

Regiment of 880 soldiers: JY 1:209.

Regiment blocked at Muxiyuan traffic circle shifts west: JY 1:190.

"A convoy of about ten military trucks": CDP-0215.

Twelve hundred soldiers stopped on West Qianmen Street: *Time* Magazine Editors, p. 38.

Troops emerge from the Great Hall at 9:00: CDP-1030, CDP-1215. Identified as members of the Palace Guard: CDP-1022.

Troops near the Beijing Hotel swing belts at foreign cameramen: CDP-0929. One of the propaganda videos made after the Massacre splices in footage of these soldiers, taken illegally from a CTV transmission, with footage of the foot soldiers who had been stopped the night before in front of the Beijing Hotel. The point of this editorial sleight of hand was to show that the troops stopped in the early hours of Saturday morning were in proper uniform and not attempting to come into the city out of uniform.

Twenty-four soldiers escorted east by student monitors: CDP-1022.

Street fights outside Zhongnanhai at 4:40: CDP-0801.

Undercover agent captured and taken to the Monument: CDP-1020.

Inflammatory speech over the loudspeakers of the Workers' Autonomous Union: Several observers independently of one another suspected infiltration when they heard this speech (e.g., CDP-1032).

Foreigners warned off the streets late Saturday: CDP-1032 (5:00 P.M. on West Changan); CDP-0215 (11:15 P.M. on Qianmen Street). Journalists also detected plainclothes officers around the Square (CDP-0801, CDP-0929).

"You should leave now": CDP-0215.

"Grab him!": CDP-1009.

The first units to the west receive orders at 5:00 P.M.: JY 1:84–86.

"Get the fuck out of here": JY 1:86.

Police attempt to clear a path for troop trucks west of Gongzhufen: Zhongyang Beijing shiwei xuanchuanbu, *Pingxi baoluan*, p. 76.

Commander of the Beijing Military Region stopped at 9:55: JY 1:93.

"We will put an end to the turmoil": JY 1:88.

"Put down your arms!": My account of fighting around the Military Museum and Muxidi is drawn substantially from the account in *Paris Match* by journalist Pierre Hurel, "Dans la nuit sanglante," pp. 75–76.

Stun grenades outside the Military Museum at 10:35: CDP-1022.

"All of sudden someone yelled": CDP-1004.

Student from Beijing University among the first people shot at Muxidi: ASA 17/60/89, p. 17.

Rumor that Deng gave written permission to use ammunition: CDP-1023. Jiang, *Countdown to Tiananmen*, pp. 319–20, says that Vice-President Wang Zhen was with the troops on West Changan and arranged for authorization to fire. The source is not reliable: Jiang was only a former instructor at a Party school and had no acccess to reliable sources at the top.

"In front of the flaming barricade": Hurel, "Dans la nuit sanglante," p. 75.

"By this time the trolley bus on the main bridge was on fire": CDP-1004.

"There was so much gunfire between us and the hospital": Adie, "Non-stop News," p. 5.

People did not believe that soldiers were shooting citizens: CDP-0716, CDP-1020.

"We understand the students' demands": CDP-0304.

APC 332: JY 2:110–16.

"Fly across the top of the Square": CDP-1030.

"A strange noise": CDP-1032.

"The APC sent people scattering like water from a boat": CDP-0718.

Unmarked white van collecting the body outside Qijiayuan: CDP-1022.

"The student organizers of the blockade decided to reinforce": CDP-1032.

"The APC hit one of the five troop trucks and flipped it": CDP-0929.

"The civilian's head was popped like a grape": CDP-0718.

"The APC had to slow down considerably to get through": CDP-1032.

"Look, your own army's killing you!": CDP-0718.

Published testimony of Colonel Dong Xigang and two crew members: JY 1:103–111. Colonel Dong has also reported his experience in the hospital: JY 2:313–16.

"Didn't know the route of attack": JY 2:179.

Eighteen APCs join the column at Fuxingmen: JY 1:90.

Two APCs fire tear gas at Xidan at 11:45 P.M.: JY 1:141.

Fighting in the Fuxingmen area: CDP-1022.

"I watched about twenty men in combat gear": ASA 17/60/89, p. 18

"The soldiers ran out of tear gas": ASA 17/60/89, p. 18.

"They ran after the trucks and shouted protest slogans": ASA 17/60/89, p. 19.

"Then, without warning, the troops opened fire on us": Wu Ming, "I Witnessed the Beijing Massacre," p. 6.

"I was told that the mass of people had blocked the army": Lu, "Beijing Diary," p. 17.

Official account of Liu Guogeng's death: Zhongyang Beijing shiwei xuanchuanbu, *Pingxi baoluan*, pp. 19–21.

Inscription on the side of the bus at Xidan: CDP-1062.

Stories of people being shot in their apartments: CDP-1090; ASA 17/60/89, p. 15.

The Army regiment coming from Marco Polo Bridge: Zongzhengzhibu xuanchuanbu, *Hanwei shehuizhuyi gongheguo*, pp. 307–312.

Identification of troops as Air Force: Lao Gui, quoted in Human Rights in China, *Children of the Dragon*, p. 144. He estimates their number at over one thousand, full regimental size.

Contingent coming from the Railway Station: JY 2:85–87.

Unit in the south end of the city receives authorization to go ahead at any cost: JY 2:203.

"Ahead of schedule": JY 1:188.

"Those of us on the Square had been hearing news": CDP-1030.

The second APC passes the Beijing Hotel at 12:50: CDP-0907.

"The screaming around me rose even louder": Simpson, "Tiananmen Square," pp. 21–23. Later that night, journalist Jan Wong asked a student monitor directing traffic around the APC what had happened to the crew. The monitor assumed the student rescue operation had been successful. "He said there were only two and they had been rescued by students. They were slightly injured and students took them away and put them in a bus. It wasn't at the stage where they were going to tear soldiers apart" (CDP-0907).

"I could see soldiers and APCs": CDP-0801.

No shooting victims in the medical tents: CDP-0718.

Text of the government broadcast at 1:10: *Time* Magazine Editors, *Massacre in Beijing*, pp. 56–57.

"The broadcast told us to go home": CDP-0907.

Special PSB team frisking reporters at the Beijing Hotel: JY 1:296 says that the team arrived at the hotel in plainclothes at 1:40, switched into uniform, then inspected every foreigner going in and out until 7:30 A.M.

"I got to the northeast corner of the Square": CDP-0805.

"Can you come? Can you come?": CDP-0801.

Person escapes by jumping into the moat: CDP-0718.

PAP waiting inside the Forbidden City: CDP-0806.

Man insists soldiers would use rubber bullets: CDP-1022.

Soldiers move from the west to cordon off the Square: CDP-0215.

"Soldiers advance shortly after 2:00": CDP-0907.

"The students lit a fire barricade": CDP-0602. This testimony is taken from a word-processed eyewitness account posted on June 7 on a booth near South Zhongguancun Road, in the university district. It is dated June 5.

Capital Iron and Steel worker counts twenty-nine dead: Yu and Harrison, *Voices from Tiananmen Square*, p. 193.

"Every few minutes, there was a burst of gunfire": CDP-0928.

"People behind us were pressing us toward the Square": CDP-0215.

"People were becoming more and more enraged": CDP-0806.

"I need to get in to rescue the wounded!": CDP-0806.

"Down with Deng Xiaoping!": CDP-0215.

Student shot in the leg at Liubukou: CDP-0314.

"Some people started screaming that they were hit by bullets": The account that follows merges the voices of the two men, CDP-0717 and CDP-0718.

Soldiers on the steps of the Museum: An hour later, Robin Munro of Asia Watch found them still there, awaiting orders. "All along the wide steps in front of the Museum, and all the way back up into the proscenium, sat row upon row of soldiers in steel helmets, several thousand of them in all, each one holding a thick wooden cudgel" (Munro, "Beijing on the Night of June 3–4," p. 5).

The Army's northward advance from south of the city: CDP-1022.

Regiment halted at the east gate of the Temple of Heaven: JY 1:209–212.

Gunfire heard in the southeast: CDP-0907 first noted the sounds of fighting to the south at 3:57. CDP-1022 observed heavy thuds in the southeast at 4:50.

"It was as though a civil war had taken place": CDP-0215.

State Council spokesman Yuan Mu on Cui Guozheng: *The June Turbulence in Beijing*, p. 15. Cui is not mentioned by name in Yuan's account. A eulogy for Cui appears in Zhongyang Beijing shiwei xuanchuanbu, *Pingxi baoluan*, pp. 6–12, though it offers no information on the circumstances of his death.

"Sometime after 3:00 A.M., three Army trucks came": ASA 17/60/89, pp. 25–26.

"Almost constant automatic gunfire": CDP-0928.

The assertion that no one was killed in the Square between 4:30 and 5:30 was made by a political commissar with the martial law forces, Zhang Gong, at a press conference on June 6: *The June Turbulence in Beijing*, p. 9.

Dead and wounded removed from the Square for the two hours before the clearing operation: CDP-0215, CDP-0805.

"There were four casualties inside": CDP-0207.

Hou Dejian's negotiations for withdrawal: Yi and Thompson, *Crisis at Tiananmen*, p. 242; JY 1:263–67.

Three thousand soldiers squatting on the steps of the History Museum: Nations, "Who Died, and Who Didn't," p. 12.

PLA commanders on the roof of the Great Hall: JY 2:240.

"Sparks flying in all directions": CDP-0602.

"Immediately the troops surrounding the Square began firing indiscriminately": Wu Ming, "I Witnessed the Beijing Massacre," p. 6.

The toppling of the Goddess of Democracy: JY1:259–62.

"A row of armored vehicles was moving very slowly": Munro, "Beijing on the Night of June 3–4," p. 7.

"If tanks were used to roll over people": *Fourth Plenary Session*, p. 14. The government was accused of allowing APCs to crush students in the Square at a memorial meeting students held at Wuhan University on the evening of June 5. The meeting was used to convict one of the student leaders in Wuhan, Li Haitao, of "the crime of spreading counter-revolutionary propaganda" (CDP-0829).

"It was a most horrible photograph": CDP-1062. A government publication that insists no students in tents were run over cites the case of three out-of-town students from Ningxia University who were presumed run over in the Square and mourned in a student ceremony; soon after, two of them turned up back home in Yinchuan and the third was found alive in Beijing: Zhou, "Shigatsu yirai," p. 21.

"Many helmeted military police, most of them PLA": CDP-0602.

Gunfire at 5:20 to flush out a sniper: JY2:230–31.

Rumor that the alleged sniper was a PAP: CDP-1023.

"She must have shouted something at a soldier": Wu Ming, "I Witnessed the Beijing Massacre," p. 7.

Student sources insist that students were killed during the evacuation of the Square: CDP-0705, CDP-0907, CDP-1090. In an interview, Wuerkaixi also understood that some students who refused to withdraw were shot, but has no hard evidence for this. Opposing testimony, that no one was shot during the withdrawal, comes from Robin Munro and Lao Gui, both quoted in Human Rights in China, *Children of the Dragon*, p. 174. Hunger striker Zhou Duo also testified that no students were shot in the southeast corner of the Square: see Lu, "Beijing Diary," p. 22.

"I noticed three PLA soldoers": Munro, "Beijing on the Night of June 3–4," p. 7.

"We didn't fight anyone, we just left": CDP-1090.

"Seven died instantly": CDP-0907.

"I had a hell of a time identifying this photo": CDP-1062.

Students seen trudging north to the campuses: CDP-1004.

Chapter 6. Counting Bodies

Shirts hung at university gates punctured with a line of holes: CDP-1062, referring to Beijing University.

"It happened so quickly and so violently": Adie, "Non-stop News," pp. 5–6.

Journalists find pools of blood, entrails, brain matter, bullet holes: CDP-0801, CDP-0805.

Soviet journalists estimate ten thousand killed: *Time* Magazine Editors, *Massacre in Beijing*, p. 69.

"Troops came in from five directions": CDP-0718.

Crowds gather at Beishida: CDP-1061, CDP-1215. At Qinghua: CDP-1024.

"Some were hysterical and couldn't stop crying": CDP-1062.

Assertion that Beishida student was run over by APC: CDP-0927.

Bodies put on display at campuses: CDP-1009, at Beijing Foreign Languages Institute; CDP-0915, at the Chinese University of Politics and Law.

Snapshots posted at People's University: CDP-0915.

Guns and clothes displayed at intersections: CDP-0201.

Estimate of three thousand given at intersection near Beishida: CDP-1090.

Estimate of three thousand given by Beishida students on campus: CDP-0201.

Estimate of three thousand given on the steps of the Beijing Hotel: CDP-0801.

Television news estimate of two thousand: CDP-0215.

Radio Beijing states "thousands" dead: FBIS-CHI-89-106, p. 94.

Student handbill declares over two thousand dead and tens of thousands wounded: A099 (June 4).

Beida Students' Autonomous Union announcement that at least twenty-one hundred died: Duke, *The Iron House*, p. 120.

Beida handbill accepts the three thousand figure: A110 (June 4).

Federation handbill estimates seven thousand wounded: translated in "Beijing Spring 1989," p. 126.

Chinese Red Cross gives figure of twenty-six hundred to CBC: Simmie and Nixon, *Tiananmen Square*, p. 194. That figure was explicitly denied on Radio Beijing on June 16 (CBB, News from Radio Beijing, Vancouver, June 18, 02:00:09 GMT).

"A few thousand": CDP-0201.

Alleged order to empty hospital beds: CDP-1063.

Alleged order to accept only cadres and soldiers: CDP-1022.

Doctor warned not to go out on the street Saturday night: CDP-1004.

Hospitals not prepared for the scale of wounded: Agence France-Presse, in *La Presse* (Montréal), 6 June 1989.

Hospitals lack sufficient medical supplies, especially blood for transfusions: Sheryl WuDunn, *New York Times*, June 5, 1989.

"Some people rushed in with a man on a stretcher": Yige Yisheng, "Yige yisheng yanli," pp. 73–74.

Beijing Hospital begins to receive wounded at 2:16 A.M.: CDP-0907. Casualty figures for the Beijing are not available.

Dead and wounded brought to the Capital Hospital at 2:00 A.M.: CDP-0805.

Bus ferrying wounded from the Square changes hands twice as drivers are shot: CDP-0806.

"The busiest time was around 5:00 A.M.": CDP-1020.

"It looked like an abattoir": Jasper Becker, London *Guardian*, June 5, 1989.

Beida student visits the Posts and Telecommunications Hospital: Duke, *The Iron House*, p. 119.

Burn victims at No. 3 Hospital: CDP-0717.

Fuxing Hospital: ASA 17/60/89, pp. 16, 17; CDP-0716; CBB, Beijing Massacre, Indiana University, June 4, 05:20:56 GMT.

Crowds gather at hospitals: CDP-0215, CDP-0929, both referring to the Capital Hospital.

Xuanwu Hospital: CDP-1090.

History major from Beida detained: CDP-0705.

Canadian Television crew at the Capital Hospital: CDP-0929.

A nurse working at the Capital Hospital reports forty deaths: CDP-1022.

Beijing doctor contacts eleven hospitals and tallies 500 dead: reported on Hong Kong Commercial Radio report (FBIS-CHI-89-106, p. 55). Other aggregate figures were given on Sunday. The Japanese broadcasting corporation NHK contacted three hospitals and came up with the figure of 80 dead. A leaflet posted at Beida counted 111 dead at nine hospitals (both ibid., p. 52). A leaflet pasted on a traffic sign reported that six hospitals had admitted about 200 dead and wounded as of 6:00 A.M. (Duke, *The Iron House*, p. 118).

"Were full of wounded and dying people": CDP-0603. According to Hong Kong's *Wen wei po*, other than those listed in Table 6–1, Tiantan, Ritan, Jingwen, and Jishuitan hospitals also received corpses (FBIS-CHI-89-107, p. 35).

Thirty-two hospitals care for hunger strikers: Zhonggang Beijing shiwei bangongting, *1989' Beijing zhizhi dongluan*, p. 108.

"This is one of those that Li Peng calls rioters": CDP-1090.

"Bandaged students up as quickly as they could": CDP-1061. The same story was reported in *Wen wei po* (FBIS-CHI-89-107, p. 36). It was rumored that soldiers did remove students from hospitals, but I have found no direct evidence for this.

Injured students transferred from Capital to No. 3 Hospital: CDP-0717. BBC journalist Kate Adie heard that some Capital doctors were shot for resisting Army orders (cited in *Time* Magazine Editors, *Massacre in Beijing*, p. 55). No hard evidence is available.

Soldier transferred from Capital to Sino-Japanese Hospital: JY 389–92.

Thirty-four soldiers in the Capital Hospital: JY 1:335.

Soldiers admitted to civilian hospitals: Eleven were admitted to No. 4 Hospital before 3:00 A.M. (JY 1:210). Of 56 soldiers admitted before midnight to Tongren Hospital, southeast of the Square, 33 needed hospitalization (Zhongyang Beijing shiwei xuanchuanbu, *Pingxi baoluan*, pp. 44–49). At Friendship Hospital in the south end of the city, 49 underwent surgery (ibid., p. 91). The Beijing Municipal Emergency Center on the east side of the Forbidden City said it had treated altogether 181 soldiers (ibid., pp. 65–68); another official source mentions that one soldier died there (JY 2:288).

Emergency medical center inside the north entrance of the Great Hall of the People: JY 1:72.

Emergency medical center in the Temple of Heaven: JY 2:336. Runs out of food and medical supplies on Tuesday: JY 2:309.

Seven nurses accompany the troops on the night of June 3: JY 1:351–52.

Nurses in locations outside the Square: JY 2:336–37.

Beijing Garrison Hospital sends out ambulance at 11:00 A.M.: JY 2:308.

Reports that amphetamines were given to soldiers: CDP-0730, CDP-0801, CDP-1022. A Cable News Network reporter broadcast the story on June 5, saying an official was his source.

Troops loading bodies onto trucks at 11:20 A.M.: CDP-1022.

Capital Iron and Steel truck collecting bodies on the 3rd Ring Road: CDP-1022.

Rumor of fighting between Capital steelworkers and soldiers: CDP-0718.

Unmarked morgue trucks: CDP-0718. Morgue trucks probably did not deliver their contents to civilian hospitals. The Hong Kong magazine *Jiushi niandai* [the nineties] reported that Beijing crematorium workers refused to burn bodies that were not positively identified, the implication being that the Army had to cremate them on its own (cited in Duke, *The Iron House*, p. 112).

"The smell was overpowering": CDP-0801.

Beida student watching fires in the Square: CDP-1030.

"The Army used bulldozers to shovel bodies into piles": A110 (June 4).

Qinghua University student sees soldiers heap body bags together at south end of the Square: "Beijing Spring 1989," p. 124. The original appeared in Hong Kong's *Wenhui bao* on June 5; it is reprinted in Han Shanbi, *Lishi de chuangshang*, p. 163.

Report of soldiers using flamethrowers against people: CDP-0201.

The invention of the fire lance: Needham, *Science in Traditional China*, pp. 39–40.

"During this time he could hear screams": ASA 17/60/89, p. 37.

Cases of detention and torture in the Workers' Culture Palace reported in the Hong Kong press: *Ming bao*, July 28, 1989, cited in ASA 17/60/89, p. 37. One case involved a middle-school student, the other a cadre, both arrested on June 5.

Prominent general's grandson spared: CDP-0207.

Allegation that soldiers used knives to execute detainees: CDP-0620.

Radiologists told to destroy X rays: CDP-0791

Yuan Mu's press conference: *The June Turbulence in Beijing*, p. 5.

The Municipal Party Committee report: "What Has Happened in Beijing" appeared in *Beijing Review*, no. 26, and is reprinted in *The June Turbulence in Beijing*, p. 48.

Chen Xitong, June 30: reprinted in Zongzhengzibu xuanchuanbu, *Hanwei shehuizhuyi gongheguo*, p. 528.

Statistics compiled by the Army Intelligence Unit: CDP-0718.

Formal charges include accusation of spreading inflated casualty figures: CDP-0829.

"None of my closest friends was shot": CDP-0716.

"I think that the casualties were in the hundreds": CDP-1024.

Chapter 7. Consequences

Hearts and minds: The phrase was used by journalist Jim Munson, CTV News, to translate what a Chinese interviewee told him on Monday: "The People's Liberation Army may win this battle, but they have lost the war for the hearts and minds of ordinary Chinese."

Large convoy coming from the east: CDP-1022, JY 1:244–50.

"The barricades were flimsy that night": CDP-0907.

Tank convoy coming down west of Beijing Hotel: CDP-1022.

Driver stoned to death: This scene has been captured on film and is replayed in the government videos. Similar stonings of drivers in disabled vehicles occurred elsewhere that morning, e.g., JY 2:143.

"I threw myself underneath a car": CDP-0215.

Tank unit arrives twenty minutes ahead of schedule: JY 1:269.

Armored convoys entering from west: A column of tanks and APCs based west of the city was ordered into motion just after 6:00 A.M. and had to push through many barricades along West Changan Boulevard before reaching the Square at 6:50 (JY 2:74–75).

Over a hundred armored vehicles in the Square by 7:00 A.M.: CDP-1030.

Japanese reporter counts ninety-six tanks and sixty-five APCs in the Square: FBIS-CHI-89-106, p. 62.

Tanks with 105- mm guns: The original Type 59 battle tanks were built with a 100-mm gun, but some were upgraded with the U.S. M68 105-mm gun, which is licensed in Israel and sold to China by the Israelis (Lee, *China's Defence Modernization*, p. 292). These are also installed on the newer export model, the Type 69 battle tank.

Thirty-two tanks dispatched from the Square to Zhongnanhai: JY 1:269–70.

"You could cross, but you couldn't stop": CDP-1090.

Unit sent to replace the guards at Chaoyangmen: JY 2:406–8.

Firing at the northeast corner of the Square until 9:00 P.M.: Cao Xinyuan, "Wo yisheng zhong," p. 10.

"Don't be afraid": CDP-0806.

Rumor of firing from helicopters: CDP-0718.

Soldiers in the Square down to a daily ration of two biscuits by Monday: JY 2:241, 258.

Ambulances remove injured soldiers at a military airfield west of the city: CDP-0907.

"Suddenly shooting started" : CDP-0801.

Convoy blocked at the College Road–4th Ring Road intersection: CDP-1009.

Military convoys speeding through Beitaipingzhuang: CDP-1090.

"10:30 P.M. Presumably I hear the tanks coming up": CDP-0608.

Columns coming into the city Sunday night: An observer on the east side of Beijing (CDP-0718) watched a convoy of sixty tanks and APCs enter the city from the east through Jianguomen at 2:00 A.M., followed by another group of thirty armored vehicles two hours later. Altogether that night, seventy-four tanks and forty-five APCs moved in from east of the city along Changan Boulevard. The troops accompanying these convoys were nervous and fired repeatedly as they drove along.

Convoys coming into the city on Monday: One of Monday's largest convoys from the east pulled into Tiananmen Square shortly before 2:00 P.M. An observer watched from the grounds of the Beijing Hotel: "We counted ten tanks, ten APCs, seventy-five trucks, most of them with troops in them, one Red Cross car, and three water tankers. It took thirty-five to forty minutes for them to pass by. People were yelling slogans at them. As they went by the Beijing Hotel, where people were hanging out of windows and balconies, they fired several rounds. It seemed as though they were firing into the air. I lay flat on the ground, trying to keep my head below a curb of about five or six inches" (CDP-0801).

Squads of armed foot soldiers deployed along flanks of truck columns: CDP-0805.

"The soldiers fired into the air": CDP-1031.

Jianguomenwai diplomatic apartments hit by bullets: CDP-1022.

"The people there claimed they were plainclothes policemen": CDP-1004.

"You bastards call yourself Chinese?": The remark appears in the official government publication, Zhongyang Beijing shiwei xuanchuanbu, *Pingxi baoluan*, p. 83.

Arrest of Wang Weilin: reported in the London *Daily Express*, June 18, 1989; see Asia Watch, *Repression in China since June*, p. 62.

Barricades at every intersection on the northern section 3rd Ring Road: A foreigner (CDP-1215) reports that "around Hepingli we were blocked by citizens and students and we had to take a detour north. Eventually we got back on the ring road, but then we were stopped by buses placed across the ring road at Anzhenqiao. There were forty or fifty people at the intersection, and no evidence of any military around. Farther west on the ring road, around Madian, students stopped us and attached a big white paper chrysanthemum to the grill of the car and then let us through. They were not using buses to block traffic there, just people, thirty or forty, to block the road. It was very orderly." CDP-0927 provided similar testimony.

Soldiers bivouacked by Airport Road: CDP-1010; CDP-1215. Other units were encamped in suburban factories further out beyond the northeast corner of the 3rd Ring Road (CDP-0917).

"They scouted the Jianguomenwai area": CDP-0717.

"There were burned-out armored vehicles, trucks, and buses": CDP-0716.

Thirty-four APCs burned at Muxidi: CDP-1090.

"The bus was already burnt out": CDP-0603.

Soldiers from abandoned APCs beaten and admitted to No. 307 Hospital: JY 1:332.

Ming pao reports commander of APC column ordered executed: FBIS-CHI-89-107, p. 21.

Burned-out convoys on approach roads into the city: They could be seen in the area of Jianguomenwai in the east, on the southeast sector of the 3rd Ring Road, on Nanyuan Road (about twenty vehicles) and East Tiantan Road (a much larger convoy) coming up from the south around the Temple of Heaven, and on Changping and College roads coming down from the north.

Mayor Chen Xitong's speech, June 30: reprinted in Zongzhengzhibu xuanchuanbu, *Hanwei shehuizhuyi gongheguo*, p. 528.

"Somebody go out and light up one of the trucks": CDP-1031.

"The Army was still there": CDP-1032

Soldiers cannibalizing breakdowns for parts: CDP-1010.

Breakdown rate of one vehicle in fifteen: CDP-1031.

"To give the government an excuse for cracking down": CDP-0730.

Supreme Court employee sees soldiers burn six armored vehicles: CDP-0806. It has been said that soldiers were seen burning Army vehicles as late as Tuesday: Lu, "Beijing Diary," p. 22.

"Look, we didn't do that": CDP-1004.

Expectation that the Army might be rewarded materially for the suppression: The PLA seemed to expect a payoff. In May 1990, an internal PLA publication,

Junshi jingji yanjiu [Military economic research], called for more than doubling defense spending over the next decade, running counter to budgetary trends of the previous decade (Associated Press, December 14, 1990).

Preparations for an Army attack at Beishida: CDP-1061.

"Thousands of people had been killed": CDP-0917.

Rumor that the PLA had occupied Qinghua University: CDP-0915.

"There was an instant of panic": CDP-0917.

"An Appeal to the International Community": A104.

"The June Third Atrocity": A099.

"It was so busy there": CDP-1062.

Students decide on Monday not to defend the campuses: CDP-0603.

"I had previously arranged an appointment": CDP-0927.

"Was extremely tense": CDP-1024.

Tear gas exploded near the main gate of Beishida: CDP-0201, CDP-0927.

Anticipation of Army occupation of Beishida on Monday at 3:00 P.M.: CDP-0201.

Soldiers on the campus of the College of Chinese Medicine: CDP-0701, CDP-0917. It was also rumored that soldiers entered the Beijing Aviation University that night (CDP-1063), but this not been confirmed.

Student heads for Tianjin by bicycle on Monday: CDP-0927.

University administrators encourage students to leave campus: CDP-0705.

Conical graves in the middle of the road: CDP-0718.

PLA weapons lost: It was also rumored that workers were arming themselves with captured PLA weaponry. Some machine guns were removed from the APCs that burned near Muxidi and emplaced in nearby yards on Sunday (CDP-0915).

Student asks Americans to get guns from their embassy: CDP-0701.

"Very frightened and confused, and young": CDP-0201.

Variant accounts of the delivery of arms to Beishida come from CDP-0927, CDP-1061, and CDP-1128. CDP-1061 also heard that the Army tried to deliver arms to other Beijing campuses, also without success.

Students on the hoods of the Army trucks holding banners: CDP-0201.

Soldiers identify themselves as members of the 40th Group Army: Yang Jianli, quoted in Human Rights in China, *Children of the Dragon*, p. 183.

Citizens tell incoming soldiers that the 38th was rebelling: JY 2:139. This official PLA account substitutes "XX" for the "38th," but that is clearly the group army intended.

Rumor early Sunday afternoon that the 38th Army was rebelling: Hou Dejian, in Yi and Thompson, *Crisis at Tiananmen*, p. 240.

"I'm a PLA man": Gao Huan, quoted in Human Rights in China, *Children of the Dragon*, p. 188.

"We the PLA did not fail you": CDP-1022.

Soldiers wearing white armbands and red armbands: CDP-1022.

Soldiers flash the number 38 on their fingers: FBIS-CHI-89-108, p. 21.

Four hundred officers and soldiers went missing: *The June Turbulence in Beijing*, p. 5.

Rumor of group armies coming from Fujian and Guangdong: CDP-0701.

Military transport planes ferry troops from Fujian: Erbaugh and Kraus, "The 1989 Democracy Movement in Fujian," p. 150.

Site of the 1990 Asian Games used as a military execution ground: CDP-1022.

This source, a military specialist, discounted, on the basis of internal inconsistencies in the story, another widespread rumor that soldiers had been executed near Marco Polo Bridge. It was widely believed that forty officers were executed in June (CDP-1023).

Officers of the 40th Group Army court-martialed: CDP-0620.

Yang Baibing's report of officers "breaching discipline": *South China Morning Post*, December 28, 1989.

Allegation that soldiers of the 27th shot soldiers of the 38th during the Massacre: Bob Nixon, Hong Kong Commercial Radio, June 6 (FBIS-CHI-89-107, p. 24).

Secondhand reports from students and foreign journalists: Wuerkaixi (CDP-0207), BBC correspondent Kate Adie ("Non-stop News," p. 5).

NATO estimate that one thousand soldiers died: quoted in Jane's, *Crisis in China*, p. 38.

"Damned old VOA stripped everything conjectural out": CDP-0928.

"Seventeen tanks and a couple of APCs arrayed themselves": CDP-1031.

Hypothesis that the tank column was placed on the Jianguomen overpass by the 27th to defend itself from attack: I have received several unconfirmed reports that the CIA was monitoring unit-to-unit radio transmissions within the PLA during this period, and that it concluded from those transmission that some units were jockeying for a fight. The CIA has not acknowledged having done so, and a U.S. embassy official in Beijing has denied that the embassy had this information. But other sources in the Beijing diplomatic world say that this information was shared with friendly embassies as each considered the advisability of evacuating its nationals.

"Study Lei Feng!": CDP-0718.

Armored column parked one kilometer east of Jianguomen: CDP-0718.

The sound of shooting in the suburbs: CDP-0730.

"There is no way this could have been conflict": CDP-0716.

"People guessed that there was conflict": CDP-1090.

"We heard what we thought was shellfire": CDP-0605.

"We could hear heavy gunfire": CDP-0915.

"The 27th Army shot at the unit coming in": CDP-0917.

Report of a U.S. marine regarding disabled armored vehicle: CDP-0927.

BBC news team in the suburbs: Adie, "Non-stop News," p. 6.

BBC and VOA reports of conflict around Nanyuan: CDP-1024.

"They'd had the hell beaten out of them": CDP-0718.

"Wishful thinking": CDP-0201.

State Council running out of gas: CDP-0718.

"Open troop carriers, each one full of soldiers": CDP-0730.

PLA units still coming into the city on Thursday: JY 2:450.

"Soldiers with straw brooms were out sweeping the streets": CDP-0901.

"When I went back out into the streets around 7:30": CDP-0901.

Police begin to round up workers on the weekend: An official publication says that the PSB conducted house-to-house searches in western Beijing on June 12 and 14, rounding up almost one hundred suspects. Charges were laid against twenty-six for attacking military vehicles and hiding military goods and ammunition. *Pingxi baoluan*, p. 77; see also JY 2:412.

"The weather has changed": CDP-0901.

Chapter 8. Closing the Century

"Never change China back into a closed country": *Fourth Plenary Session*, p. 18.

"Figured that the worst that could come of this": CDP-1090.

"We knew that something was going to happen": CDP-1061.

"The longer the wait went on": CDP-1063.

"They announced that a counter-revolutionary incident had been put down": CDP-0907.

Plainclothes officers relieving barricaded soldiers: for example, on Saturday evening, a squad of plainclothes police posing as ordinary citizens surrounded a military jeep trapped near the railway bridge south of Yongdingmen and were able to protect the three occupants from harm (Zhongyang Beijing shiwei xuanchuanbu, *Pingxi baoluan*, p. 40).

Foreign military observers did not detect agitators in the crowds: CDP-1022.

"The more people there were in the Square": CDP-0201.

Teams of marksmen used on June 3: JY 2:137.

British-made traffic monitors: The tapes from these monitors are imprinted with date and time. The first time one of the government videos was shown, it included a traffic monitor tape whose time imprint directly contradicted the voice-over narration. This imprint was removed from subsequent copies.

"China would surely have been thrown into serious turmoil": "Beijing fengbo jishi." The voice-over is in English.

"Did the soldiers open fire first": CDP-0705.

"When the Chinese government repressed its minorities": Jigme Ngapo, "The Lesson of the Beijing Massacre," p. 13.

"As long as the demonstrations were absolutely peaceful": CDP-1063.

"This would mean the destruction, overnight": FBIS-CHI-89-115, p. 10. Yang is citing remarks made by Chen Yun, which he repeated on May 26; cf. Oksenberg et al., *Beijing Spring, 1989*, p. 332.

"That is the part that is revolting": CDP-1022.

Soldiers seen with garottes and steel-core whips: CDP-1020.

Soldier animated by a subjective commitment to sacrifice himself before harming the people: for example, JY 2:141–42.

Army reports involvement of units from eighteen provinces: JY 2:156, in which it says that units from twenty provinces and municipalities were involved. To my knowledge, Beijing and Tianjin were the only contributing municipalites.

"We must protect the army of our friends": JY 2:269.

First major supply convoy gets into the city Tuesday night: JY 2:259.

Soldiers sent out to shops for batteries: JY 2:269.

"The soldiers seemed to be under pretty good control": CDP-0730.

Commanders and commissars of six military regions relieved of office or transferred in May 1990: *Asahi shimbun*, June 14, 1990.

Officers of the 64th at Beidaihe; representative of the National Education Commission at the National Defense University: CDP-0314.

Chinese press report in April 1989 that 85 percent of eighteen-year-olds applied to join the PLA: *China Daily*, April 29, 1989, citing *Zhongguo gingnian bao* [China Youth News], reprinted in FBIS-CHI-89-083, p. 114. The article makes direct reference to the problem posed by the low educational level of recruits.

"This was the first time in history": CDP-0716.

"This kid was constantly on the phone": CDP-1021.

"Even conservative-minded intellectuals said": CDP-1024.

"Civil society": This concept has been applied to the Democracy Movement by Lawrence Sullivan in "The Emergence of Civil Society in China" and by David Strand in "Protest in Beijing." For observations regarding the weakness of civil society in China at the time of the Massacre, see Esherick and Wasserstrom, "Acting Out Democracy," pp. 858–60.

North China Herald: February 14, 1900. I am grateful to my student Greg Patterson for sharing with me this and the following passage.

"Every siege in history has been like this": B. L. Putnam Weale, *Indiscreet Letters from Peking* (New York: Dodd Mead, 1919), p. 154.

"We are Chinese": Jy 2:139.

"To guard against the subversive and destructive activities": Guo Jisi, "Oppose the Use of Human Rights Issues," p. 22.

Nonrecognition of the rights of the individual in Chinese law: These observations derive from a discussion with William Jones, whose ideas regarding Qing law have influenced me greatly.

"A criminal in the scales of history": CDP-0604.

"Reverse the verdict on the Tiananmen incident yourself": CDP-0716.

"Members of the Beijing Party Committee used the Movement": CDP-0618.

Bibliography

Publications in Western Languages

Adie, Kate. "Non-stop News." *The Listener*, September 21, 1989, pp. 4–7.

Amnesty International. "People's Republic of China: Preliminary Findings on Killings of Unarmed Civilians, Arbitrary Arrests and Summary Executions since 3 June 1989." August 14, 1989. ASA 17/60/89.

———. "The People's Republic of China: Trials and Punishments since 1989." April 1991. ASA 17/34/91.

Asia Watch. *Punishment Season: Human Rights in China after Martial Law.* New York: Asia Watch, 1990.

———. *Repression in China since June 4, 1989: Cumulative Data.* New York: Asia Watch, 1990.

———. "Update on Arrests in China." November 15, 1989.

"Beijing Spring 1989: The People's Democracy Movement, Documents from the Participants." *Our Generation* 21:1 (September 1989): 100–132.

Duke, Michael. *The Iron House: A Memoir of the Chinese Democracy Movement and the Tiananmen Massacre.* Layton, Utah: Gibbs Smith, 1990.

Erbaugh, Mary S. and Richard Kurt Kraus. "The 1989 Democracy Movement in Fujian and Its Aftermath." *Australian Journal of Chinese Affairs*, no. 23 (January 1990): 145–60.

Esherick, Joseph, and Jeffrey Wasserstrom. "Acting Out Democracy: Political Theater in Modern China." *Journal of Asian Studies* 49:4 (November 1990): 835–65.

The Eyes Have It. Chinese title: *Liu-si jianzheng.* 2d ed. Hong Kong, 1990. Text in both English and Chinese.

Fathers, Michael, and Andrew Higgins. *Tiananmen: The Rape of Peking*. London: Independent/Doubleday, 1989.

Foran, Charles. "Beijing Voices." *Descant*, no. 70 (Fall 1990): 51–62.

Fourth Plenary Session of the CPC 13th Central Committee. Beijing: New Star Publishers, 1989. Includes Deng Xiaoping's June 9 talk.

Guo, Jisi. "Oppose the Use of Human Rights Issues as an Excuse to Interfere in Interal Affairs." Translated in *Human Rights Tribune* 1:5 (1990): 19–23. Chinese original published in *People's Daily*, overseas edition, January 1990.

Han, Minzhu, ed. *Cries for Democracy: Writings and Speeches from the 1989 Chinese Democracy Movement*. Princeton: Princeton University Press, 1990.

Holland, Gill. "Beijing Spring." *Alumni Magazine of Washington and Lee* 64:4 (December 1989): 17–20, 37–39.

Human Rights in China, Inc., ed. *Children of the Dragon: The Story of Tiananmen Square*. New York: Collier, 1990.

Hurel, Pierre. "Dans la nuit sanglante de Pékin." *Paris Match*, June 15, 1989, pp. 75–76.

Jane's Information Group. *China in Crisis: The Role of the Military*. Coulsdon, Surrey: Jane's Information Group, 1989.

Jiang, Zhifeng. *Countdown to Tiananmen: The View at the Top*. San Francisco: Democratic China Books, 1990.

Jochnowitz, George. "True Power to the People." *National Review*, June 30, 1989, pp. 22–23.

The June Turbulence in Beijing. Beijing: New Star Publishers, 1989.

Kristof, Nicholas. "China Update: How the Hardliners Won." *New York Times Magazine*, November 12, 1989, pp. 66–68.

Landsberger, Stefan R. "Chronology of the 1989 Student Demonstrations." In *The Chinese People's Movement: Perspectives on Spring 1989*, ed. Tony Saich, pp. 164–89. Armonk, N.Y.: M.E. Sharpe, 1990.

Lee, Ngok. *China's Defence Modernization and Military Leadership*. Sydney: Australian National University Press, 1989.

Li, Kwok Sing. "Deng Xiaoping and the Second Field Army." *China Review* (January 1990): 40–41.

Lipman, Jonathan N., and Stevan Harrell, eds. *Violence in China: Essays in Culture and Counterculture*. Albany: State University of New York Press, 1990.

Lu, Yuan. "Beijing Diary." *New Left Review*, no. 177 (September-October 1989): 3–26.

Ma, Shu Yun "The Rise and Fall of Neo-Authoritarianism in China." *China Information* 5:3 (Winter 1990–91): 1–18.

Martin, Helmut. *China's Democracy Movement 1989: A Selected Bibliography of Chinese Source Materials*. Köln: Bundesinstitut für Ostwissenschaftliche und Internationale Studien, 1990.

Munier, Bruno. *Voyage au Printemps de Pékin*. Montréal: Méridien, 1989.

Munro, Robin. "Beijing on the Night of June 3–4: A First Hand Account." *Human Rights Watch*, no. 3 (September 1989): 1, 5–7.

Nations, Richard. "Who Died, and Who Didn't." *Spectator*, July 29, 1989, pp. 11–13.

Needham, Joseph. *Science in Traditional China*. Cambridge: Harvard University Press, 1981.

Ngapo, Jigme. "The Lesson of the Beijing Massacre." *News Tibet* 23:2 (August 1989).

Oksenberg, Michael; Lawrence Sullivan; and Marc Lambert, eds. *Beijing Spring, 1989: Confrontation and Conflict: The Basic Documents.* Armonk, N.Y.: M. E. Sharpe, 1990.

Rolph, Hammond. "The PLA Army in 1987: The Long Road to Modernization." In *SCPS Yearbook on PLA Affairs, 1987*, pp. 51–69. Kaohsiung: Sun Yat-sen Center for Policy Studies, 1988.

Saich, Tony, ed. *The Chinese People's Movement: Perspectives on Spring 1989.* Armonk, N.Y.: M. E. Sharpe, 1990.

Shapiro, Fred C. "Letter from Beijing." *New Yorker*, May 26, 1989, pp. 73–82.

Simmie, Scott, and Bob Nixon. *Tiananmen Square.* Vancouver: Douglas and McIntyre, 1989.

Simpson, John. "Tiananmen Square." *Granta*, no. 28 (Autumn 1989): 9–25.

Strand, David. "Protest in Beijing: Civil Society and Public Sphere in China." *Problems of Communism*, 39:3 (May–June, 1990): 1–19.

Sullivan, Lawrence R. "The Emergence of Civil Society in China, Spring 1989." In *The Chinese People's Movement: Perspectives in Spring 1989*, ed. Tony Saich, pp. 126–44. Armonk, N.Y.: M.E. Sharpe, 1990.

Testimonial to History. Chinese title: *Lishi de jianzheng.* Hong Kong: Hong Kong Alliance in Support of Patriotic Democratic Movements of China, 1989. Text in both Chinese and English.

The Peking Massacre: A Summary Report of the 1989 Democracy Movement in Mainland China. Taipei: Kwang Hwa Publishing Company, 1989.

The Truth about the Beijing Turmoil. Chinese title: *Beijing fengbo jishi.* Beijing: Beijing Chubanshe, August 1989. Text in both Chinese and English.

Time Magazine Editors. *Massacre in Beijing: China's Struggle for Democracy.* New York: Warner Books, 1989.

Wakeman, Frederic. "The June Fourth Movement in China." *Items* (Social Science Research Council, New York) 43:3 (September 1989): 57–64.

Wasserstrom, Jeffrey, and Elizabeth Perry, eds. *Popular Protest and Political Culture in Modern China: Learning from 1989.* Boulder, Colo.: Westview Press, 1991.

Watson, Burton. *Basic Writings of Mo Tzu, Hsün Tzu, and Han Fei Tzu.* New York: Columbia University Press, 1967.

Witness Reports on the Democratic Movement of China '89. Chinese title: *Bajiu Zhongguo minyun jianzheng baogao zhuanji.* Hong Kong, 1990. Text in both English and Chinese.

Wu Ming [pseud.]. "I Witnessed the Beijing Massacre." *China Update*, no. 3 (December 1990): 6–7.

Young, Robert. *The Dragon's Teeth: Inside China's Armed Forces.* London: Hutchinson, 1987.

Yu, Mok Chiu, and J. Frank Harrison, eds. *Voices from Tiananmen Square.* Montréal: Black Rose Books, 1990.

Publications in Asian Languages

Bajiu Zhongguo minyun baozhang touban zhuanji [Newspaper front pages on the Democratic Movement of China in 1989]. Hong Kong: Zhongguo Minzhu Yundong Ziliao Zhongxin, 1989.

Beijing fengbo jishi. English title: *The Truth about the Beijing Turmoil*. Beijing: Beijing Chubanshe, August 1989. Text in both Chinese and English.

Cao Xinyuan. "Wo yisheng zhong zui changde sishi xiaoshi" [The longest forty hours of my life]. *Zhongguo zhi chun* [China spring], no. 75 (August 1989): 8–10.

Ding Chu. *Minzhu yundong de Sun zi bingfa* [Master Sun's art of war in the Democracy Movement]. San Francisco: Minzhu Zhongguo Shulin Chubanshe, 1990.

Guojia jiaowei sixiang zhengzhi gongzuosi [Ideological and political work bureau of the National Education Commission]. *Jingxin dongpo de wushiliu tian* [Fifty-six shocking days]. Beijing: Dadi Chubanshe, August 1989. For internal circulation.

Han Shanbi, ed. *Lishi de chuangshang: 1989 Zhongguo minyun shiliao huibian* [The wounds of history: collected historical materials on the 1989 China Democracy Movement]. 2 vols. Hong Kong: Dongxi Wenhua Shiye Gongsi, 1989.

"Lujun budui bingli ji peizhi zhuangkuang" [Army strength and disposition]. *Zhongguo dalu* [Mainland China monthly] 22:7 (July 1989): 34–35.

Quanguo yiyuan minglu: gongye ji qita bumen ji jiti suoyouzhi yiyuan [National list of hospitals: hospitals owned by industries and other departments, and collectively owned hospitals]. Beijing: Renmin Weisheng Chubanshe, 1987.

"Tiananmen guangchang baowei buzhang—Xin Ku" [Xin Ku, head of the Security Department of Tiananmen Square]. *Zhongguo zhi chun*, no. 76 (September 1989).

Wu Mouren et al., eds. *Bajiu Zhongguo minyun jishi* [Daily reports on the movement for democracy in China, April 15–June 24, 1989]. 2 vols. N.p., n.d. [1989].

Xingtian and Yiye, eds. *Liangci Tiananmen shijian* [The two Tiananmen incidents]. Hong Kong: Tianhe Chuban Jigou, 1989.

Yige Yisheng [A doctor]. "Yige yisheng yanli de beican zhi ye" [A tragic night in the eyes of a doctor]. *Zhongguo zhi chun*, no. 77 (October 1989): 73–74.

Zhonggong Beijing shiwei bangongting [Administrative office of the Beijing Municipal Communist Party Committee], ed. *1989' Beijing zhizhi dongluan pingxi fangeming baoluan jishi* [Record of stopping the 1989 Beijing turbulence and quelling the counterrevolutionary turmoil]. Beijing: Beijing Ribao Chubanshe, 1989.

Zhongyang Beijing shiwei xuanchuanbu [Propaganda department of the Beijing Municipal Party Committee]. *Pingxi baoluan xianjin shiji xuanbian* [Outstanding events during the suppression of the turmoil]. Beijing: Yinshua Gongye Chubanshe, July 1989.

Zhou Wenhua. "Shigatsu yirai no Pekin no keisei" [The situation in Beijing since April]. *Jinmin Chugoku* [People's China], 1989,9 (September 1989): 18–29.

Zongzhengzhibu wenhuabu [Cultural affairs department of the PLA General Political Department]. *Jieyan yiri* [A day of martial law]. 2 vols. Beijing: Jiefangjun Wenyi Chubanshe, 1989.

Zongzhengzhibu xuanchuanbu [Propaganda department of the PLA General Political Department] and Jiefangjun bao bianjibu [Editorial department of the Liberation Army Daily]. *Hanwei shehuizhuyi gongheguo* [Defending the socialist republic]. Beijing: Changzheng Chubanshe, 1989.

Videos

"Beijing fengbo jishi." English title: "A Record of the June Turbulence in Beijing." Voice-over in English. Distributed by the China Audio and Video Corporation. July 1989.

"Gongheguo weishi" [Guardians of the republic]. Produced by the PLA August First Film Studio. Beijing, 1989.

"Piaoyang, gongheguo de qizhi" [Flutter aloft, flag of the republic]. Produced by the Propaganda Section of the PLA General Political Department. Beijing, 1989.

Leaflets and Newssheets

CHINA DOCUMENTATION PROJECT, UNIVERSITY OF TORONTO

A001 "Zhu shou!!!" [Didn't lift a hand!!!]. April 20.

A002 "Shikan jinri zhi Zhongguo, jing shi shei zhi tianxia" [Take a look at today's China: whose empire is this?]. N.d. [April].

A004 "Xinwen daobao" [News herald], no. 2. May 4.

A025 "Zhongguo renmin jiefangjun jieyan budui zhihuibu gao Beijing shi shimin shu" [Statement of the Martial Law Headquarters of the People's Liberation Army to the citizens of Beijing municipality]. May 21.

A026 "Junfang qiwei gaoji jiangling biaotai" [View of seven high-ranking army generals]. N.d. [handed out on May 23]. C:R-2.

A027 "Jianjue zhizhi dongluan de xuanchuan kouhao" [Propaganda slogans for the suppression of turmoil]. May 22.

A028 "Zhi jiefangjun zhizhanyuan" [To PLA commanders]. Issued by Beida. N.d. [May 21]. C:R-6.

A031 "Wu-erling can'an zhenxiang" [The true picture of the May 20th atrocity]. Issued by Beida. May 21.

A039 "Weixian de fagui bu ju falü xiaoli" [Unconstitutional regulations have no legal force]. Issued by Beida. N.d. [May 22?].

A040 "Yiwei lao junren zhi xuesheng de gongkaixin" [An open letter to the students from a former soldier]. May 22. C:R-3.

A050 "Youxing kouhao" [Slogans for demonstrations]. Issued by Beida. N.d. [May 23?].

A051 "Gao jiefangjun zhizhanyuan shu" [Statement to PLA commanders]. Issued by Beida. N.d. [May 23?].

A057 "Zhi fengming jinzhu Beijing jieyan de guanbing de xin" [Letter to martial law officers and soldiers ordered into Beijing]. Signed "An Infantry Colonel." Issued by Beida. N.d. [May 23]. C:R-10.

A062 "Minxin bu ke ru" [The people's hearts may not be insulted]. Issued by Beida. May 24.

A063 "Yonggande zhanqilai, gongren laodage" [Stand up bravely, brother workers]. Issued by Beida. N.d. [May 24].

A064   "Jinji huyu shu" [An urgent call]. Issued by the University of Politics and Law. N.d. [May 24?].

A065   "Dangjin Beijing ren de gongde yishi" [The moral consciousness of Beijing people these days]. Issued by Beida. N.d. [May 25?].

A069   "Baozhi zhaiyao" [Newspaper digest]. Issued by Beida. N.d. [May 27?].

A071   "Shoudu gejie lianxi huiyi guanyu shiju de shengming" [Declaration of the joint conference of the various circles in the capital regarding the present situation]. May 27.

A072   "Xinwen daobao" [News herald], no. 6. May 27.

A074   "Gongheguo jue bushi siyou caichan" [The republic is not private property]. Signed Chen Wen. Issued by Beida. May 28. Reprinted in the June 3 edition of *Xinwen daobao* (A094).

A077   "Beigaolian jianbao (3)" [Brief report of the Beijing Autonomous Students' Federation, no. 3]. Signed Wang Youcai. May 29.

A079   "Zhongyangtai zemmale?' [What's up with Central Broadcasting?]. Issued by Beida. May 29.

A085   "Qunzhong laigao" [Letters from the masses]. Issued by Beida. May 31.

A091   "Xuanchuanbu kuaixun: Nanjing changtu" [Bulletin from the Propaganda Department: long distance call from Nanjing]. Issued by Beida. June 2.

A093   "Bu min zhi ye muji ji" [Eyewitness record of a sleepless night]. Issued by Beida. June 3.

A094   "Xinwen daobao" [News herald], no. 8. June 3.

A096   "Xuanchuanbu kuaixun" [Propaganda department bulletin]. Issued by Beida. June 3.

A097   "Xinwen baodao" [News report]. Issued by Beida. June 3.

A098   "Xinwen (liu-san xiawu)" [News, June 3, afternoon]. Issued by Beida. June 3.

A099   "Liu-san can'an" [The June 3rd atrocity]. June 3.

A100   "Kuaixun" [Bulletin]. Issued by Beida. June 4.

A101   "Xinwen daobao" [News herald], *haowai* [special edition]. June 4.

A104   "Dui guoji shehui de huyu!" [An appeal to the international community!]. Handwritten. N.d. [June 4].

A105   "Kuaixun" [Bulletin]. Issued by Beida. June 4.

A109   Untitled: Handwritten casualty list as of 3:00 P.M. Sunday. June 4.

A110   "Gao quan shijie tongbao shu" [Statement to our compatriots all over the world]. Issued by Beida. June 4.

STARR LIBRARY, COLUMBIA UNIVERSITY

C:R-0   "Kuaixun" [Bulletin]. May 15.

C:R-5   "Gao jiefangjun guanbing shu" [Statement to the officers and soldiers of the PLA]. May 20.

C:R-7   "Gao renmin jiefangjun quanti guanbing shu" [Statement to all officers and soldiers of the People's Liberation Army]. N.d.

C:R-8   "Gao zidibing shu" [Statement to our brother soldiers]. Issued by Beijing Aviation University. May.

C:R-9   "Ji sanshiba jun Xu junzhang bei jiezhi de jingguo" [Events surrounding the dismissal of 38th Group Army Commander Xu]. May 29.

Index